The 3ds max™ 4
Quick Reference

MICHELE BOUSQUET

The 3ds max™ 4
Quick Reference

autodesk Press

THOMSON

LEARNING

Australia · Canada · Mexico · Singapore · Spain · United Kingdom · United States

THOMSON
★
LEARNING

The 3ds max™ 4 Quick Reference
Michele Bousquet

Business Unit Director:
Alar Elken

Executive Editor:
Sandy Clark

Acquisitions Editor:
James DeVoe

Development Editor:
John Fisher

Editorial Assistant:
Jasmine Hartman

Executive Marketing Manager:
Maura Theriault

Channel Manager:
Mary Johnson

Marketing Coordinator:
Karen Smith

Executive Production Manager:
Mary Ellen Black

Production Manager:
Larry Main

Production Editor:
Tom Stover

Art/Design Coordinator:
Mary Beth Vought

Cover art by Phillip Prahl

Library of Congress Cataloging-in-Publication Data

ISBN 0-7668-3888-9

NOTICE TO THE READER

CONTENTS

CHAPTER 2 Modeling

CHAPTER 3 Lights & Cameras

CHAPTER 5 Animation

APPENDIX How to Use the CD

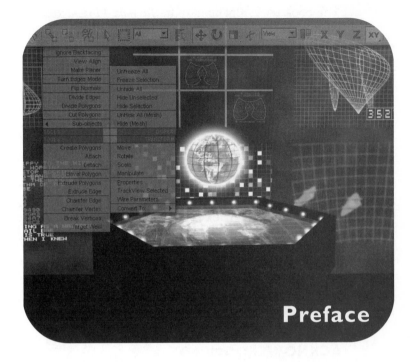

Preface

3ds max 4 is the fourth release in Autodesk's 3D Studio MAX line of software. Up until this release, the software was known as 3D Studio MAX. Recently, industry veteran Discreet took over development and marketing of 3D Studio MAX, and the software's name was changed to **3ds max 4** to better fit with Discreet's existing line of editing and compositing software. Discreet is a division of Autodesk, Inc.

3ds max 4 is a PC-based software package that allows you to model objects in three dimensions, as if you were sculpting on the computer. You can then add color and lighting, then place a camera to shoot the scene from any angle. All objects, including the camera, can be animated.

With **3ds max 4**, you can make 3D animation for videos and websites, still images for print, and backgrounds and characters for game environments. **3ds max 4** is also ideal for creating 3D elements for seamless compositing with live action. Depending on the artist's intent, the resulting images can be made to resemble anything from a stylized cartoon to a photorealistic environment indistinguishable from the real thing.

HOW TO USE THIS BOOK

This book is a companion to the *3ds max 4 Interactive Quick Reference CD*. The CD contains a comprehensive reference on every single option in **3ds max 4** along with other features. For more information on how to install and use the CD, see *Appendix: How to Use the CD*.

The information in this book is designed to get you started with **3ds max 4** and provide tips and guidelines for your development as a 3D artist. Each chapter after Chapter 1 has a *Quick Start*, and sections on *Workflow, New Features in 3ds max 4* and *Tips and Techniques*.

If you're new to **3ds max 4**, do the *Quick Start* tutorials, then explore the CD to find additional tools for developing 3D scenes. There are many more features in **3ds max 4** than those described in the *Quick Start* exercises. The *Quick Start* guides will give you enough familiarity with each feature to enable you to use the CD effectively.

The *Workflow* section is designed to help you develop efficient work practices. New and experienced users alike will find useful information in these sections.

If you've used **3ds max** before, see the *New Features in 3ds max 4* and *Tips and Techniques* sections in each chapter. The information contained in these sections is geared toward experienced users.

ACKNOWLEDGMENTS

I would like to thank the following persons for their help with this book and the accompanying CD.

Matthew T. Rodgers and Aaron Ross for technical editing and assistance

Jenny Good for copy editing

Matthew T. Rodgers for contributions to CD content

Michael George for CD compilation and programming

And, of course, the staff at Autodesk Press for helping to make this idea a reality.

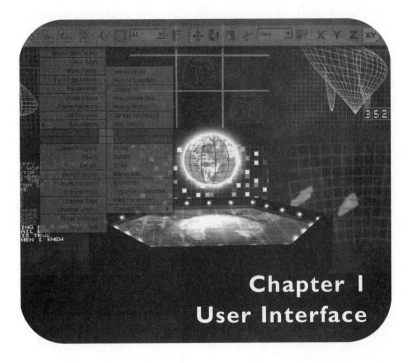

Chapter 1
User Interface

The user interface is the command control for all your activities in **3ds max 4.** Taking a few moments to learn the names and uses of each part of the interface (UI) will make it easier for you to follow tutorials and instructions in this book and other references.

USER INTERFACE ELEMENTS

The default user interface for **3ds max 4** opens automatically when the program is loaded. There are five main parts to the user interface:

- Menus across the top of the screen

- Main Toolbar just below the menus

- Command panel down the right of the screen

- Four viewports labeled Top, Front, Left and Perspective taking up the bulk of the screen

- Other controls across the bottom of the screen, including the status bar, snap tools, animation controls and view controls

There are numerous ways to customize the user interface. However, the default interface works just fine for most applications. Use the default interface setup until you get a feel for how you'd like the program to work for you, then change the setup to suit your needs. See *Customizing the User Interface* on page 7.

If your user interface doesn't look like the figures in the pages that follow, choose *Customize/Load Custom UI Scheme* from the menu and choose **DefaultUI** from the file selector.

MENUS

The menus appear across the top of the screen. Click on any menu to display a full list of options.

Figure 1.1 *Menus*

Many menu options are also available as Main Toolbar buttons. For example, the *Edit/Align* menu option can also be accessed by clicking the **Align** button on the Main Toolbar.

MAIN TOOLBAR

The Main Toolbar is displayed just under the menus.

Figure 1.2 *Main Toolbar*

If your screen resolution is less than 1280x1024, you won't be able to see all the Main Toolbar buttons at once. To access the buttons at the rightmost end of the Main Toolbar, move the cursor over the toolbar until a small hand cursor appears, then click and drag to the left.

When you move the cursor over a button, a tooltip appears after a few moments with the name of the button.

The Main Toolbar contains three tools that you will use frequently.

Select and Move

Select and Rotate

Select and Uniform Scale

These tools are called *transforms* because they transform objects in one way or another.

Flyouts

 When a button image has a small triangle at its lower right corner, it can be held down for a moment to display more buttons. This stack of buttons is called a *flyout*. The original button appears on the flyout along with the other options.

As you move the cursor over each option in the flyout, the button name appears on the status bar at the bottom of the screen.

Once a flyout button is selected, it replaces the original button on the Main Toolbar.

COMMAND PANELS

Command panels reside on the right side of the screen.

Figure 1.3 *Command panels*

There are six command panels that can be displayed by clicking on the appropriate button at the top of the panel:

 Create for creating objects, shapes, lights, cameras and other scene elements

Modify for modifying objects already created

Hierarchy for managing hierarchies of linked objects

Motion for controlling animation

Display for determining how elements display in viewports

Utilities for additional utilities for working with objects and scenes

Command panels are divided into sections called *rollouts*. A rollout has a title with a (+) or (-) sign at the left end of the title. You can roll a rollout up or down like a window-shade by clicking its title. The **Name and Color** rollout is shown in Figure 1.3.

VIEWPORTS

The four default viewports take up the largest part of the screen. Viewports allow you to view and work on the scene from a variety of angles. Each viewport has a *label*, such as Top, Front, Left or Perspective.

Objects, lights, cameras and other scene elements are created and positioned in viewports. Once a camera has been placed in a scene a viewport can be made to show what the camera sees. Any viewport, including a camera view, can be rendered.

Resizing Viewports

You can change the size of any viewport by clicking and dragging on the crossbars that separate the viewports. This feature is new in **3ds max 4**.

Viewport Pop-up Menu

Right-clicking on the viewport label accesses a pop-up menu with a variety of options for working with viewports.

Figure 1.4 *Viewport right-click pop-up menu*

The *Smooth & Highlights* and *Wireframe* options change the way objects are displayed in viewports. The *Views* option selects from different types of views such as Front and Perspective.

Quad Menu

Right-clicking in any viewport brings up the *quad menu*. This menu features a variety of commands which vary according to the operation currently being performed. The quad menu is so named because it usually consists of four quarters. Under some circumstances, only two of the four quarters will appear when the quad menu is accessed.

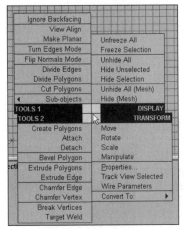

Figure 1.5 *Quad menu*

The quad menu can be customized to display specific commands.

OTHER CONTROLS

Across the bottom of the screen are several areas for viewing information or controlling the scene.

Status Area

The status area displays information related to the current activity.

The pink and white areas displays text used when writing MAXScripts.

Figure 1.6 *Status area and transform type-in*

Immediately to the right of the pink and white areas is the prompt line. The top line displays the current selection status, while the lower line gives hints or directions about the next action to take.

The transform type-in area consists of three entry fields labeled X, Y and Z. These fields display the position, rotation or scale of the currently selected object. You can also type numbers directly into these fields to transform an object with the currently selected transform.

Toggles

The toggle buttons turn specific features on and off.

Figure 1.7 *Toggles*

Four of these buttons feature a magnet as part of the button image. These are the *snap tools*. When a snap tool is turned on, it causes values to change by a specific increment. For example, turning on the **Angle Snap Toggle** causes any rotation of objects to occur at 5-degree increments.

Time Controls

The time controls at the bottom right of the screen are used to work with animation.

The Animate button enables you to animate objects by changing them in the scene.

Many of the time control buttons resemble those on a VCR. Use these buttons to play or stop animation onscreen, or to move the scene forward or backward in time. You can also type in a specific frame number to jump directly to that frame.

Figure 1.8 *Time controls*

Viewport Navigation Controls

The viewport navigation controls allow you to zoom, pan or rotate the view in one or all viewports.

Figure 1.9 *Viewport navigation controls*

When a camera viewport is activated, the viewport navigation controls change to display options specifically for moving the camera and/or target.

Figure 1.10 *Controls for a camera viewport*

CUSTOMIZING THE USER INTERFACE

You can customize just about every aspect of the **3ds max 4** user interface. You can also save any number of custom setups for later retrieval.

All customization of the UI takes place on the Customize User Interface dialog. To access this dialog, choose *Customize/Customize User Interface* from the menu.

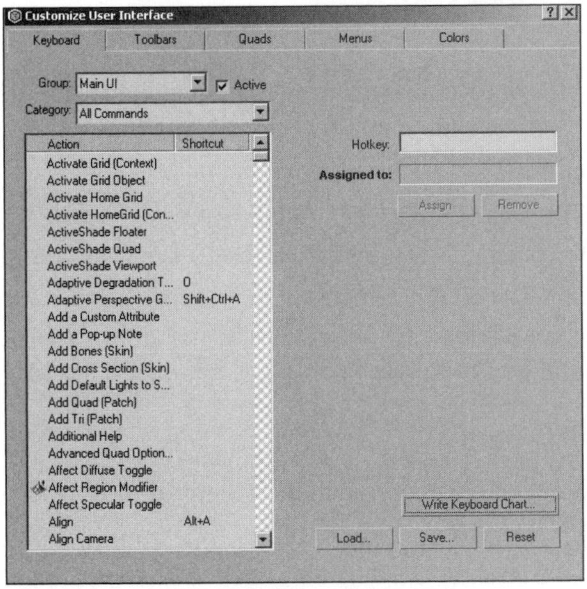

Figure 1.11 *Customize User Interface dialog*

The colors of viewport backgrounds, selected objects, and numerous other screen elements can be customized.

COLORS

The colors of most UI elements can be customized. To perform this customization, click the **Colors** tab on the Customize User Interface dialog.

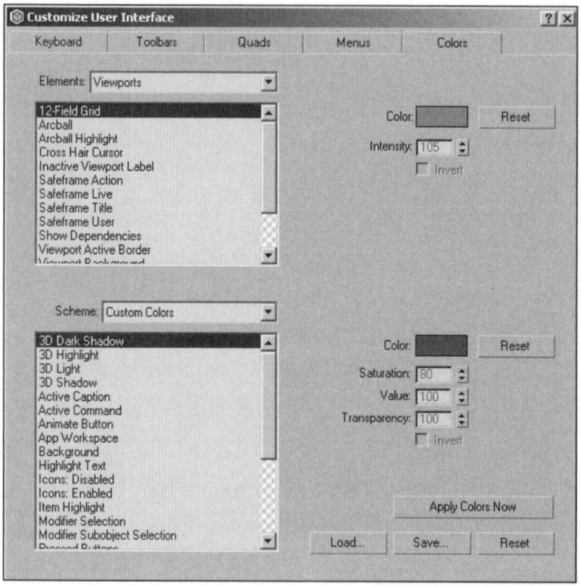

Figure 1.12 *Colors tab*

Choose a category from the **Elements** pulldown menu, then highlight a choice from the list just below Elements. Click the **Color** swatch at the right to display a Color Selector, and change the color for the selected element.

To see your change take place immediately, click **Apply Colors Now**. Otherwise, your changes will go into effect when you close the dialog.

TOOLBARS

You can create your own custom toolbars for specific uses. In addition, toolbars can be locked to any outer edge of the viewport area, or can float free on the screen.

Docking and Floating Toolbars

A toolbar is either *docked* (locked to the top, front, left or right of viewports) or *floating* (moving freely in the viewport area).

To dock or float a toolbar, right-click at the left or top end of the toolbar, where the double bars appear. On the pop-up menu that appears, choose *Dock* then a screen area, or choose *Float*.

A floating toolbar can be quickly docked by dragging and releasing it near a docking area.

Creating a Custom Toolbar

On the Customize User Interface dialog, click the **Toolbars** tab. Click **New**, and enter a name for your custom toolbar. The toolbar appears as a small box on the screen.

Choose an option from the **Groups** pulldown menu. Locate the option you would like to have on your toolbar, and drag it from the list to the toolbar. If an option already has a button image associated with it, the button image appears on the toolbar. If not, the option name appears on the toolbar.

To change the button image or option name to a specific button image, right-click the option on the toolbar and choose *Edit Button Appearance* from the pop-up menu. The Edit Macro Button dialog appears.

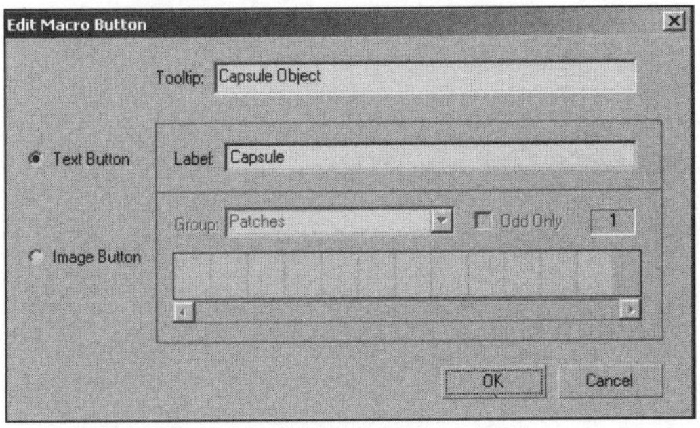

Figure 1.13 *Edit Macro Button dialog*

Choose the **Image Button** option. Making a selection from the **Group** pulldown menu will display a series of button images. Click the desired button image and click **OK**. The new button image appears on the toolbar.

Tab Panel

A *tab panel* consists of several tabs containing command sets in button form. Any toolbar can be made to be a tab panel, and any tab panel can be turned into a toolbar.

To display the tab panel, right-click any non-button area of a toolbar to display a pop-up menu, and check the *Tab Panel* option.

SAVING AND LOADING A CUSTOM UI

To save a custom setup of colors and toolbars, click **Save** on the **Customize User Interface** dialog. then enter a name for your custom UI. You can also save the current user interface by choosing *Customize/Save Custom UI Scheme* from the menu.

To load a custom UI in the future, choose *Customize/Load Custom UI Scheme* from the menu.

Take care not to save over the file DefaultUI, which contains the default UI that comes with **3ds max 4**.

NEW IN 3DS MAX 4

Two important changes have been made to the UI in **3ds max 4.**

Viewports can be resized interactively by clicking and dragging on the dividing bars that separate viewports.

A new feature, the *quad menu*, has been added to **3ds max 4**. The quad menu appears when you right-click in a viewport. See *Quad Menu* on page 5.

SUMMARY

The user interface (UI) consists of five main parts: menus, the Main Toolbar, command panels, viewports, and additional controls across the bottom of the screen.

It's important to know the names of each area so you can follow tutorials and instructions in this book and other books.

Most aspects of the UI can be customized. All customization operations are performed from the *Customize* menu, or by right-clicking on toolbars and tab panels.

For additional information on customizing the user interface, see the *3ds max 4 Interactive Quick Reference CD*.

Chapter 2
Modeling

Modeling is the process of creating 3D objects. Many primitive 3D objects can be made directly with simple commands in **3ds max 4**. In addition, **3ds max 4** has many tools for modeling and modifying complex objects.

All modeling starts with a primitive 3D object or a few shapes. Once the basic object is created, modifiers can be used to modify, or change, the simple object into a more complex object.

Additional tools on the Create panel can be used to create complex objects such as character models.

QUICK START

Like most new users, you are probably eager to get started with **3ds max 4**. This Quick Start section is designed to give you a jump-start so you can begin to experiment and learn.

Creating a Sphere

1. Load **3ds max 4**.

   The default **3ds max 4** user interface appears.

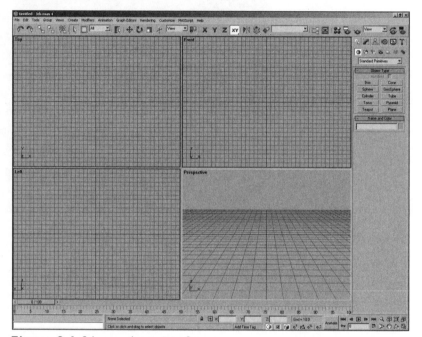

Figure 2.1 *3ds max 4 user interface*

Four viewports are displayed, labeled Top, Front, Left and Perspective. Down the right side of the screen is the command panel. By default, the Create panel is selected, and the Geometry button is selected at the top of the Create panel.

Several options are displayed on the command panel, including Box, Cone and Sphere.

2. On the command panel, click the **Sphere** button.

3. Move your cursor near the center of the Top viewport. Click and drag to create a sphere a little smaller than the viewport, then release to complete creation of the sphere.

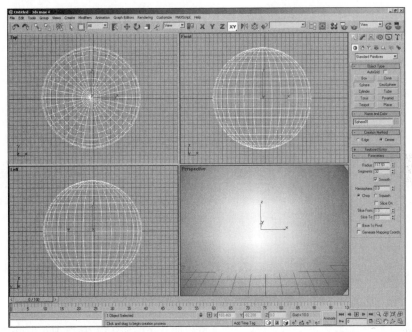

Figure 2.2 *Sphere*

You have just created a 3D object in **3ds max 4**. Note that the Top viewport is now surrounded by a yellow border, indicating that this is the active viewport.

Zooming

 In the Perspective view, the sphere is larger than the viewport. At the lower right of the screen are eight buttons known as the view controls. These buttons change the view in one or more viewports. You will use a view control button to change all views so they show the entire sphere.

 1. Locate the Zoom Extents All button in the view controls area at the lower right of the screen. Click **Zoom Extents All**.

The viewports zoom in or out to show the entire extents of the sphere.

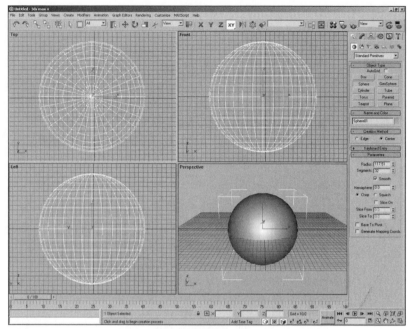

Figure 2.3 *Viewports zoomed to extents of sphere*

2. Click **Zoom All** in the view controls area.

3. In any viewport, click and drag downward to zoom out from the scene.

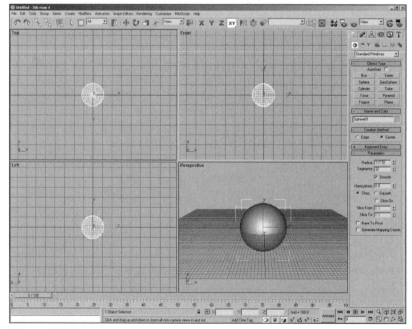

Figure 2.4 *Viewports zoomed out*

Creating the Tabletop and Teapot

Next you will create a simple scene of a teapot and sphere sitting on a tabletop. The scene shown below will give you an idea of what your final scene will look like, although your own work may differ somewhat from the scene shown.

Figure 2.5 *Simple scene*

First you will create a flat cylinder for the tabletop.

1. On the command panel, click **Cylinder**.

2. In the Top viewport, click and drag near the center of the viewport to create the top radius of the cylinder. Release the mouse.

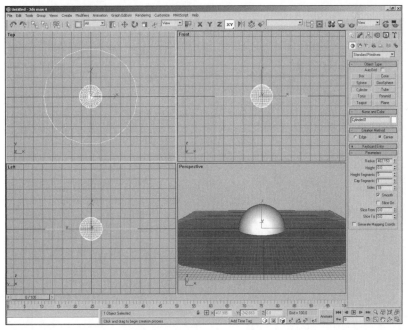

Figure 2.6 *Radius of cylinder*

3. Move the cursor downward by a small amount to create the height of the cylinder, then click to set the height.

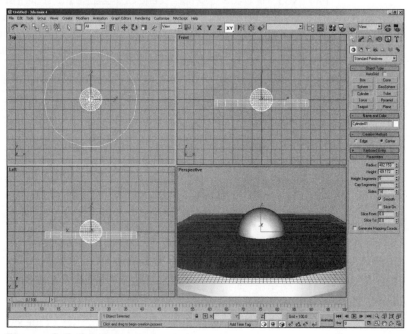

Figure 2.7 *Cylinder created*

If your table is the wrong size, just press **<Delete>** on your keyboard to delete it, and try again.

4. On the command panel, click **Teapot**.

5. In the Top viewport, click and drag anywhere in the viewport to create a teapot about the same size as the sphere.

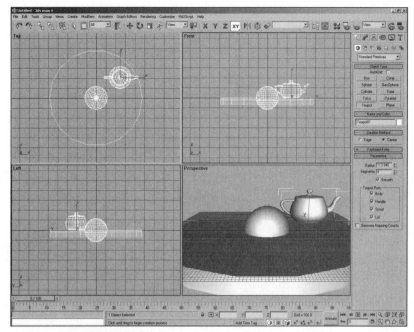

Figure 2.8 *Teapot in scene*

Moving the Objects

Now you have three objects in the scene: a cylinder, a teapot and a sphere. Next you'll arrange the teapot and sphere so they sit directly on the table.

1. Locate the Main Toolbar at the top of the screen.

   The Main Toolbar contains numerous buttons to help you create your scene.

Figure 2.9 *Main Toolbar*

2. Near the center of the Main Toolbar, locate the Select and Move button. Click **Select and Move**.

3. In the Front viewport, move the cursor over the sphere until the selection cursor appears. Click on the sphere.

   The sphere turns white to indicate that it has been selected. Three axis arrows are also displayed on the sphere.

4. In the Front viewport, click and drag the sphere upward to move it just above the tabletop.

   You can click and drag on any one of the axis arrows to move the object in just the direction specified by the arrow.

5. Move the teapot in the same way so it sits just above the table.

6. In the Top viewport, move the teapot and sphere as necessary to arrange them on the table.

 7. Click **Zoom Extents All**.

☀ **TIP** ☀

The Perspective view is rarely used for moving and placing objects. The Top, Front and Left views are much more useful for exact placement.

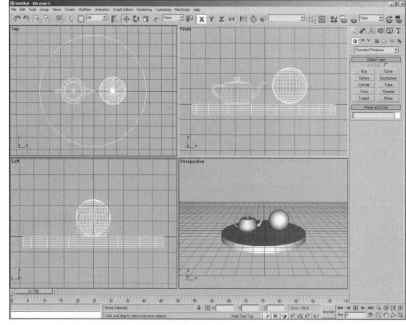

Figure 2.10 *Objects arranged on tabletop*

Resizing Objects

Suppose you find that the sphere is too big, and you would like to decrease its radius. This can be accomplished by changing the sphere's parameters on the Modify panel.

1. Click on the sphere to select it.

 2. Click the **Modify** button at the top of the command panel.

The Modify panel displays the parameters for the sphere. Next to each numerical parameter are a set of up and down arrows. These arrows, called *spinners*, can be used to interactively adjust the parameter.

3. Place your cursor directly over either of the spinner arrows next to the **Radius** parameter. Click and drag downward.

The size of the sphere decreases.

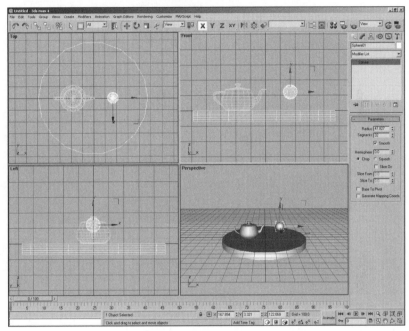

Figure 2.11 *Sphere decreased in size*

4. Move the newly resized sphere so it sits on the tabletop.

Now that you know how to create, move and resize primitives, you can add a leg and base to the table.

5. Create two cylinders for a leg and base for the table, and move them into place. Your scene should look similar to the Figure 2.12.

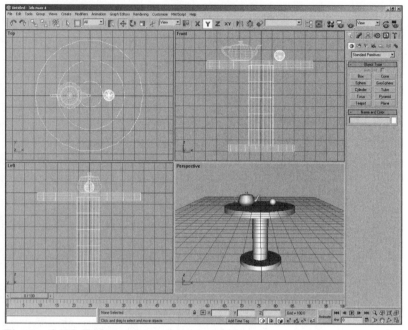

Figure 2.12 *Table with base*

Rendering the Scene

After the scene has been set up, it can be rendered.

1. Click in the **Perspective** viewport to make it the active viewport.

2. Use the **Zoom** and **Pan** buttons to arrange the Perspective view to look similar to Figure 2.12.

3. Locate the Quick Render (Production) button at the right end of the Main Toolbar. Click **Quick Render (Production)**.

A new window opens, and the Perspective view is rendered with a black background.

☙ **TIP** ☙

If your screen resolution is less than 1280x1024, you won't be able to see the Quick Render (Production) button on the Main Toolbar. To access the buttons at the rightmost end of the Main Toolbar, move the cursor over the toolbar until a small hand cursor appears, then click and drag to the left.

Figure 2.13 *Rendered image with black background*

Changing the Background Color

By default, **3ds max 4** renders all images with a black background. Here you'll learn how to change the background color.

1. On the *Rendering* menu, choose *Environment*.

   The Environment dialog appears.

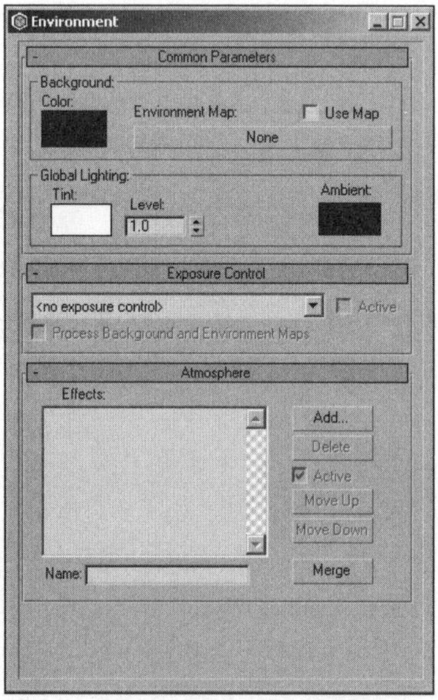

Figure 2.14 *Environment dialog*

2. Click the color swatch under **Color** at the upper left of the dialog.

   The Color Selector dialog appears.

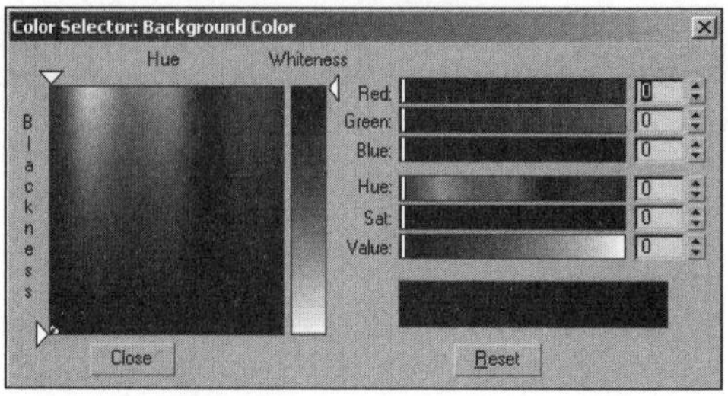

Figure 2.15 *Color Selector*

3. Change the color to white by pulling the **Whiteness** slider to the bottom.

4. Click **Close** to close the Color Selector.

   The rendering background color has been changed to white. Note that the viewport background colors have not changed -- the color change will appear only in renderings.

5. Close the Environment dialog by clicking the **X** at its upper right corner.

 6. Click **Quick Render (Production)**.

The scene renders with a white background.

Figure 2.16 *Rendered image with white background*

Adjusting a Viewport

Before placing more objects in the scene, we'll customize the Front viewport for easier modeling.

1. Click in the **Front** viewport to activate it.

2. Press the **<G>** key on your keyboard to turn off the grid in the Front viewport.

   In the next set of steps you'll be modeling in a small area of the Front viewport, and your job will be easier if the work area is larger on the screen. As a first step, you'll use the Min/Max Toggle to cause the Front viewport to increase in size and cover the entire viewport area.

3. In the view controls area at the bottom right of the screen, click **Min/Max Toggle**.

   You can also use Region Zoom to zoom into a specific area of a viewport.

4. In the view controls area, click **Region Zoom**.

5. In the Front viewport, click and drag to create a bounding area containing the upper right third of the viewport, as shown below.

Figure 2.17 *Bounding area in Top viewport*

The view will zoom to the area you just selected.

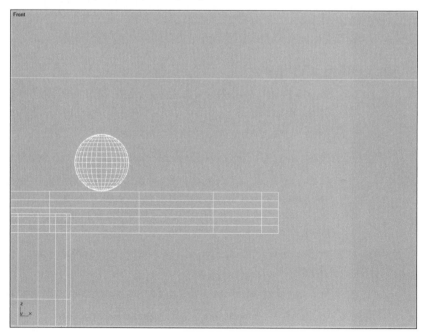

Figure 2.18 *Zoomed view of Top viewport*

Creating a Shape

You can also create 3D objects by starting with a 2D (flat) shape and giving it depth in a number of ways. Here you'll create a 2D shape that will be turned into a 3D vase.

1. On the Create panel, click **Shapes**.

2. Click **Line**.

3. In the Front viewport, draw a line similar to the one shown in Figure 2.19. Click to create the topmost point, then move the cursor and click at each point to create the line. When you have finished placing points, right-click to finish the line.

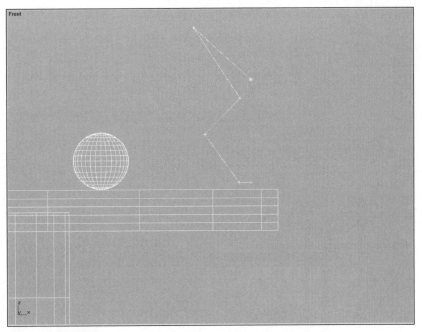

Figure 2.19 *Line*

A point on a line is called a vertex. Next, you will smooth out the line by working with its vertices.

Modifying a Shape

1. Click the **Modify** button at the top of the panel area to access the Modify panel.

   The Line object appears on the modifier stack display. The *modifier stack* lists the base object and any modifiers applied to the object in reverse order. No modifiers have been applied yet, so the base object, Line, is the only item on the stack.

2. Click the **[+]** button to the left of the Line object on the modifier list to expand the listing.

   Three items appear below the Line object: Vertex, Segment and Spline. These are the sub-objects that make up a line.

3. Click on the **Vertex** item to select it.

4. In the viewport, click and drag to draw a bounding area around the four topmost vertices on the line.

   This will select the four vertices within the bounding area. Vertices turn red when selected.

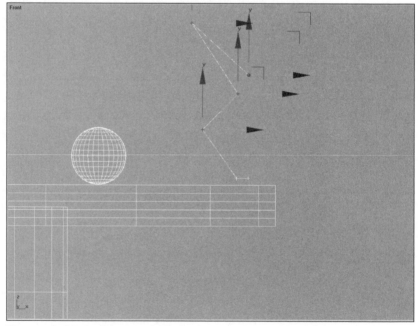

Figure 2.20 *Four vertices selected*

5. Right-click on any selected vertex. The quad menu appears.

   The quad menu is a four-part menu that appears around the cursor location when you right-click in a viewport. It contains numerous commands for working with the current scene.

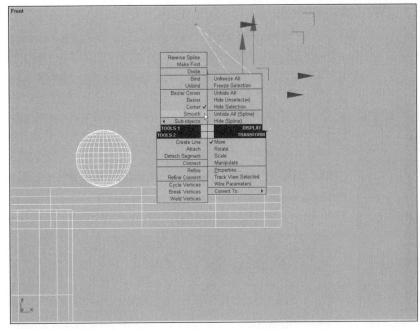

Figure 2.21 *Quad menu*

6. In the upper left quadrant of the menu, choose the **Smooth** option.

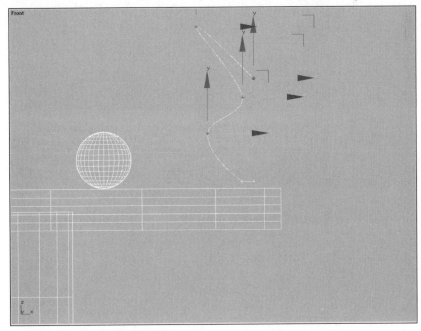

Figure 2.22 *Smoothed vertices*

The segments around the selected vertices are now smoothed out.

Applying a Modifier

A modifier can be applied to an object to change it in some way. Here, you will apply the Lathe modifier to the line shape to turn it into a 3D vase.

1. Click on the **Line** listing on the Modify panel to reselect the entire object (rather than just the Vertex sub-object level).

2. On the Modify panel, click the **Modifier List** pull-down arrow. A list of modifiers appears.

3. Click the **Lathe** modifier.

The Lathe modifier appears at the top of the modifier stack display, above Line.

In the Front viewport, you can see that the Lathe modifier has caused the shape to spin around its center., creating the shape of the vase.

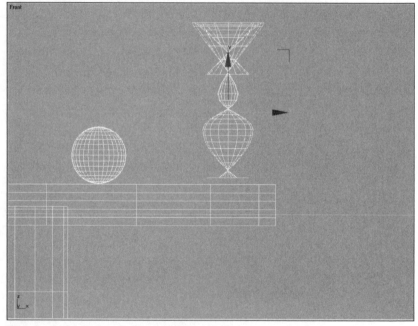

Figure 2.23 *Lathe modifier applied to shape*

Some adjustment is necessary to make the vase look right.

4. On the Modify panel, under the Align section, click the **Max** button.

The Max button causes the center axis of the lathed object to jump to the right-most side of the shape as viewed in the Front viewport. Now the vase looks correct.

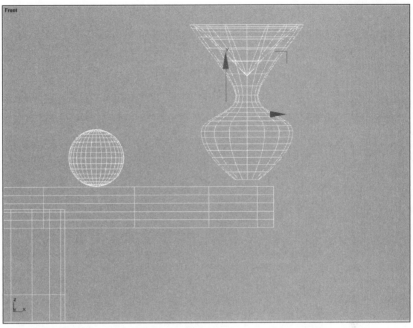

Figure 2.24 *Vase with Max axis alignment*

Now you can delete the sphere and move the vase into place.

5. Click on the sphere to select it.

6. Press the **<Delete>** key on your keyboard to delete the sphere.

 7. Click the **Min/Max Toggle** to return to the four-viewport display.

8. Click **Select and Move** on the Main Toolbar.

9. Move the vase in the Top, Front and/or Left viewports so it sits on the table.

If you position the vase correctly in the Top, Front and Left viewports first, it will automatically be correctly positioned in the Perspective view.

Applying a Bend Modifier

So far you have seen how a modifier can be used to turn a 2D shape into a 3D object. A modifier can also be used to change a 3D object itself. Here you'll use the Bend modifier to bend an ordinary cylinder into a flower stem.

1. Click the **Create** panel button to access the Create panel, then click **Geometry**.

2. Click **Cylinder**.

3. In the Top viewport, create a tall, thin cylinder to be used as a flower stem for the vase.

4. In the Front and Left viewports, move the cylinder so it sits inside the vase.

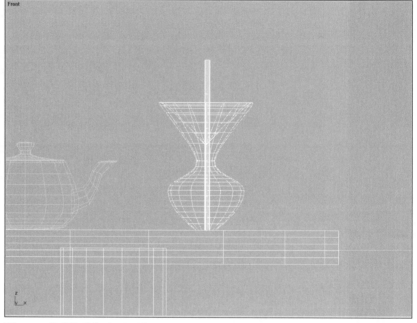

Figure 2.25 *Cylinder inside vase*

5. Click the **Modify** button to access the Modify panel.

6. If the cylinder is too wide or short, change the **Radius** or **Height** parameters to adjust the cylinder so it has approximately the correct proportions for a flower stem.

7. On the Modify panel, click the **Modifier List** pull-down arrow. Choose the **Bend** modifier.

8. On the command panel, change the **Angle** parameter to a value between 15 and 25 to make the stem bend slightly out of the vase.

The cylinder bends much like a flower stem.

Figure 2.26 *Bent cylinder*

Once the cylinder is bent, you might find that you need to move it slightly to prevent it from passing through the vase.

Creating an Extended Primitive

Extended primitives are additional 3D objects that are useful for specific needs.

1. Click the **Create** button on the command panel.

2. From the pull-down menu, choose **Extended Primitives**.

3. Click **Hedra**.

   A hedra is a mathematical object that also makes an excellent flower or ornament.

4. In the Top viewport, click and drag to create a hedra object about half the size of the vase.

5. Click the **Modify** panel button.

6. In the Family section of the panel, choose the **Star1** option.

   The hedra has been turned into a flower-like object.

7. Click **Select and Move** on the Main Toolbar.

8. In the Top, Front and/or Left viewports, move the hedra so it sits over the top end of the stem.

Figure 2.27 *Flower at end of stem*

9. To render the scene, click in the **Perspective** viewport to activate it, then click the **Quick Render (Production)** button on the Main Toolbar.

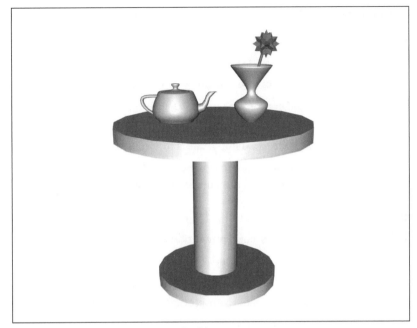

Figure 2.28 *Rendered image of table scene*

Saving the Scene

In order to load the scene the next time you load **3ds max 4**, the scene must be saved.

1. On the *File* menu, choose *Save*.

The Save File As dialog appears.

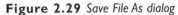

Figure 2.29 *Save File As dialog*

☀ **TIP** ☀

All scenes saved from 3ds max 4 are saved with the extension .max.

2. Enter a filename such as **Tabletop**, and click **Save**.

The next time you choose File/Save from the menu, the file will be saved with the filename you just entered.

Summary

In this Quick Start, you have learned that:

- Objects are created on the Create panel.

- Objects can be created as primitives or as 2D shapes with modifiers.

- Modifiers can be used to change the shape of an object or to turn a 2D shape into a 3D object.

- Zooming is an important and frequent part of the 3D creation process.

- To place objects correctly, they should be moved in the Top, Front or Left viewports and not the Perspective view.

WORKFLOW

No matter what the project, your modeling activities should begin with planning.

GATHER REFERENCE MATERIALS

Before beginning to model, you should have sketches or photographs of the object you plan to model. These reference materials are essential to good modeling.

DECIDE ON AN APPROACH

Every object, no matter what it is, starts out as a relatively simple 3D object of some kind. After the object is created, modifiers can be applied to it to change its shape in a variety of ways.

When deciding on an approach to modeling a particular object, look at its overall shape. Is it cylindrical, round, box-shaped? Is it tall and skinny, or short and stout? Does it have appendages or limbs? The answers to these questions will help you determine which primitives or shapes to use as a base for your final object.

CREATE BASIC OBJECTS OR SHAPES

Using your reference materials, start working on the basic model. It can be helpful to scan a reference photo or drawing and place it in the viewport background. You can do this by choosing *Views/Viewport Background* from the menu, and clicking **Files** on the Viewport Background dialog to select the file. See the *Viewport Background* topic on the *3ds max 4 Interactive Quick Reference CD* for further details.

USE MODIFIERS

Once the basic object has been created, modifiers can help you push it into shape. Familiarize yourself with as many modifiers as you can so you'll know the tools you're working with.

NEW FEATURES IN 3DS MAX 4

HOSE PRIMITIVE

The Hose primitive is a new type of Extended Primitive. This option creates a bendable accordion-style tube. You can bind the ends of the hose to two other objects. When the objects are moved, the ends of the hose go along with them.

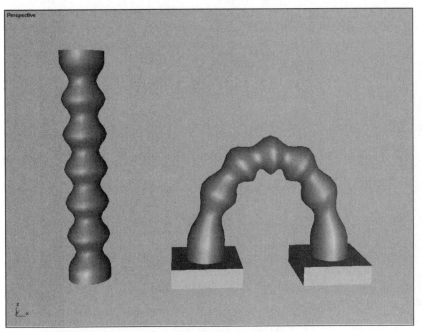

Figure 2.30 *Hose before and after being bound to objects*

HSDS MODIFIER

The HSDS modifier subdivides faces in selected parts of a mesh to give more detail. For example, you might find that the eye socket area of a facial model requires more detail.

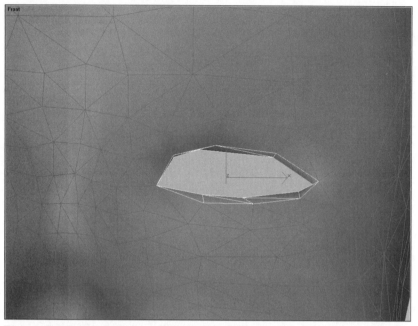

Figure2.31 *Eye of facial model with edges selected*

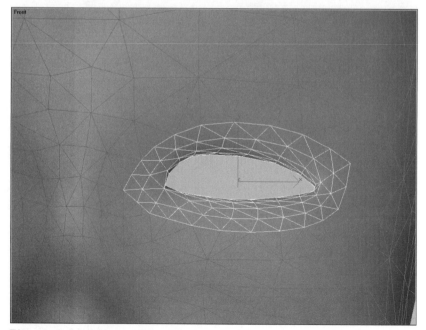

Figure 2.32 *Additional faces added by subdivision with HSDS modifier*

The HSDS modifier adds detail to just the selected edges rather than the entire mesh. Each subdivision is part of a hierarchy of levels, and you edit subdivisions at any level. The name HSDS stands for *hierarchical subdivision surfaces*.

EDITABLE POLY

A new type of object, Editable Poly, has been added to **3ds max 4**. An Editable Poly is very similar to an Editable Mesh with one important difference. Unlike an Editable Mesh, an Editable Poly's polygons can have more than three sides.

Because of its structure, an Editable Poly has a few features that an Editable Mesh doesn't have. For example, on both an Editable Mesh and Editable Poly, you can extrude selected polygons. With an Editable Mesh, all polygons extrude as one mass.

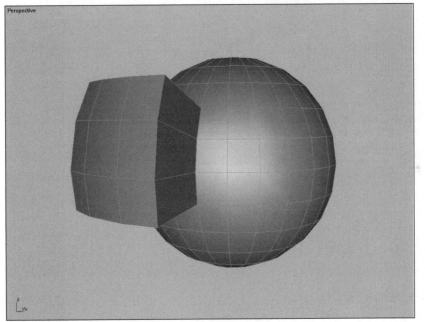

Figure 2.33 *Polygons extruded with Editable Mesh*

An Editable Poly's polygons can be extruded individually as shown in Figure 2-34.

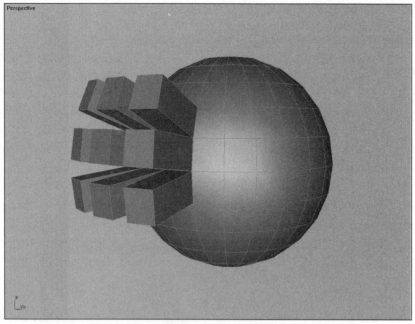

Figure 2.34 *Polygons extruded with Editable Poly*

In addition, Editable Polys have a Border sub-object that can be used to select connected open edges.

The new Poly Select modifier performs the same function for an Editable Poly as Mesh Select does for Editable Meshes and other mesh objects.

TIPS AND TECHNIQUES

Here are a few tips to help you become an expert modeler with **3ds max 4**.

KNOW YOUR TOOLS

Knowing the various 3D modeling tools in **3ds max 4** is the most important step in your progress as a modeler. Just as a successful carpenter knows which tool to use for the job, the productive 3D artist knows the software's features well enough to pick the right one at the right time.

Naturally, when you first start using **3ds max 4**, you won't know all the tools. One of the best ways to learn about them is to try them out on a simple project. After doing a few tutorials to get a feel for the software, choose a relatively simple household object, and model it in a variety of ways. A TV remote control device or small desk lamp are good choices. Place the object on your desk so you can see it easily while you model. Try to create the object with primitives, lofting, box modeling, and even NURBS if you're feeling adventurous.

If you're new to 3D, many of your early efforts might not look exactly right. However, along the way you'll learn the tools you need to make more complex objects. An approach that many artists have found successful is to alternate between these creative trials and the tutorials in the manual or a third-party book. In this way, you can learn about features you might not have found on your own, and also learn to think with the software and create by yourself.

As your familiarity with the tools grows, you'll find that you can look at any object in life and figure out how to make it with **3ds max 4**.

USE REFERENCE MATERIALS

Before setting out to create a 3D object in **3ds max 4**, you should have a strong idea of how you want the final object to look. At least one reference for the object, such as a sketch or photograph, is essential to good modeling. An exception can be made for experimentation time, where you're trying out new tools or new ways of using familiar features to get an idea of how to approach a project.

If your model "doesn't look right," ask yourself this question: What reference materials did you look at during modeling? If the answer is "None," save yourself a lot of time by getting photographs of the object or sketching it on a piece of paper. Specific differences between your 3D model and the reference picture will become obvious, making it possible for you to locate and fix the exact problem on the model.

New artists often resist using reference materials, perhaps thinking that "real" artists don't need them. Nothing could be farther from the truth. Professional artists use reference material for every project. If you doubt this, just ask the next working artist you meet.

CHOOSE THE RIGHT LEVEL OF DETAIL

You will also need to know how much detail is required on the final object. The term *detail* refers to the number of segments and/or faces on the object, which in turn determine the object's smoothness. More detail is required if the object will be viewed close up, but more detail also adds more rendering time. Ideally, you should have just enough detail to make the object look good with the camera angles you plan to use.

This underlines the need for proper planning. A modeler needs to know camera angles in advance in order to work most efficiently.

USE MORE THAN ONE APPROACH

When setting out to create a particular object, you will often find that a number of approaches come to mind (but only if you know your tools). For most objects, two or three modeling methods are possible. The one you choose depends largely on your preference and ability. Quite often, one approach is just as good as another.

It's not uncommon to start out with one approach, then reach a point where you discover that it just won't work. This is the time to switch gears and try something else. Even experienced modelers run into this situation from time to time.

Chapter 3
Lights & Cameras

The average person thinks about lighting only when the room becomes too dark to see. Conversely, when making 3D scenes, you need consider every aspect of lighting in the entire scene. Much like a photographer, you're responsible for placing the lights and setting their intensities in a way that will enhance your renderings, setting a mood or telling a story.

Lighting is sometimes given too little thought by 3D artists. If the modeling takes two weeks, then the lighting can be done in just a few hours, the reasoning goes. Too often, this isn't the case. If the scene needs a particular mood or a specific object highlighted, lighting can be the most important aspect of the scene. In addition, special effects like smoke or water always require special attention to lighting.

Camera angles and animation also contribute to the feel or mood of the scene. Although you can get satisfactory results with standard camera angles and settings in **3ds max 4,** understanding the camera options will help you use the camera for dramatic sweeps and turns when the need arises.

This quick tutorial will get you started with placing lights and cameras.

Setting up the Objects

1. Reset **3ds max 4**.

2. Create a simple scene with two objects sitting on a box.

   The box will serve as a tabletop. The Top viewport should look straight down at the tabletop, while the Perspective view should give an angled view.

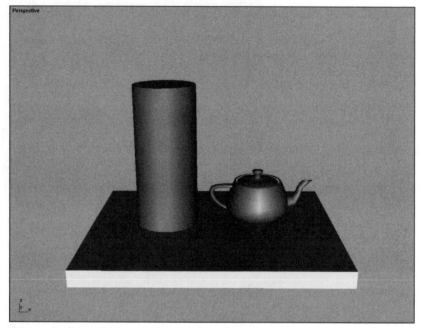

Figure 3.1 *Simple scene*

Placing the First Light

1. Zoom out the **Top** viewport to give you room to place the camera and lights.

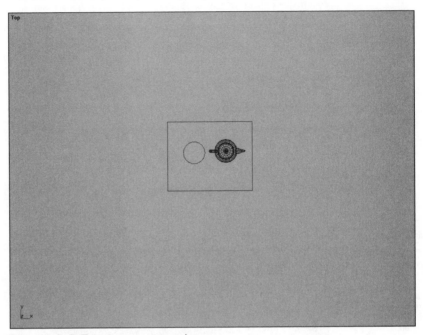

Figure 3.2 *Top viewport zoomed out*

2. On the Create panel, click **Lights**.

3. Click **Target Spot.**

4. Move your cursor to the lower right corner of the Top viewport.

5. Click and drag toward the center of the scene, and release the mouse when you reach the center of the objects.

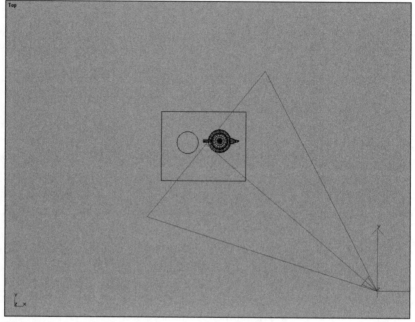

TIP

The Perspective view will go dark after you place the light in the scene. This is a normal occurrence due to the default lights turning off.

Figure 3.3 *Target Spot light placed in Top viewport*

The default lights have just turned off, leaving just the spot light in the scene.

 6. Click **Zoom Extents All**.

The light is now visible in all viewports.

The spot light consists of two parts: the light itself, and its target.

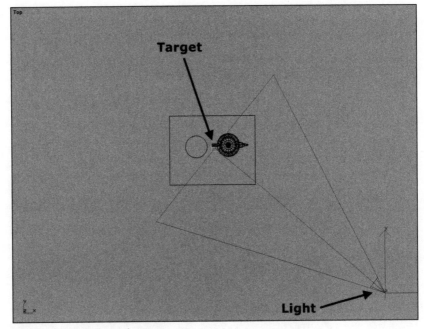

Figure 3.4 *Light and target*

In the Front and Left viewports, you can see that the light is lying flat on the construction plane. The light must be moved upward to shine on the objects from above.

Moving the Spot Light

1. Click **Select and Move**.

2. In the Front or Left viewport, click the light object (not the target) and move it upward.

   The light now shines down on the scene.

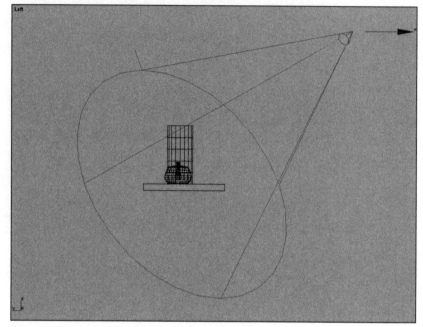

Figure 3.5 *Light moved up in Left viewport*

Placing the Camera

1. Click **Create** to access the Create panel, then click **Cameras**.

2. Click **Target**.

3. Move your cursor to the bottom center of the Top viewport. Click and drag toward the center of the scene, and release the mouse when you reach the center of the objects.

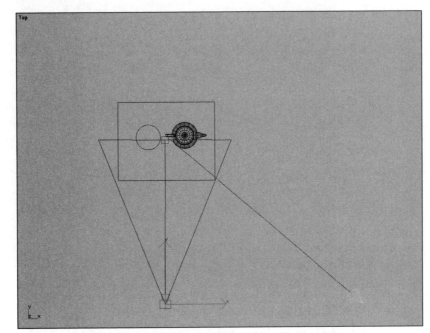

Figure 3.6 *Camera placed in Top viewport*

Like the spot light, the camera will have to be moved upward in another viewport.

4. In the Front or Left viewport, move the camera (not its target) upward, as shown in Figure 3.7.

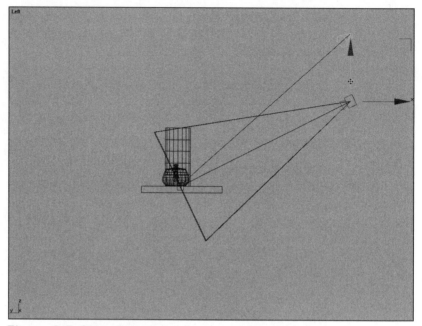

Figure 3.7 *Camera moved up*

The camera is automatically named Camera01. Next you will change the Perspective view to the Camera01 view.

5. Activate the Perspective view and press the **<C>** key on the keyboard.

6. Adjust the camera until the Camera01 view looks similar to Figure 3.8.

☙ **TIP** ☙

To quickly move both the camera and target, move the line between the camera and its target.

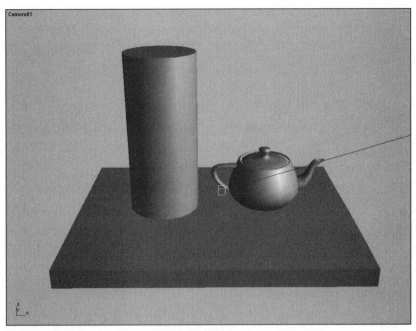

Figure 3.8 *Camera01 viewport adjusted*

 7. Activate the Camera01 view and click **Quick Render (Production)** to render the scene.

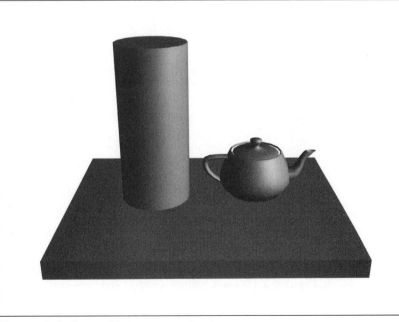

Figure 3.9 *Camera01 viewport rendered*

Because there is only one light in the scene, parts of the scene that are directly struck by the light are illuminated while the opposite sides of the objects are black. The rendering is shown here with a white background for clarity.

Next you'll balance out the lighting with a second light.

Placing a Fill Light

In photographic lighting (upon which most 3D lighting techniques are based), a light placed to fill in the dark spots is called a *fill light*. Omni lights make excellent fill lights for any scene.

1. Zoom out the Top viewport to give you room to place another light.

2. On the Create panel, click **Omni**.

3. Click once at the lower left corner of the Top viewport to place the omni light.

4. Click **Zoom Extents All**.

   This zooms all viewports except the Camera view, which always shows what the camera sees.

5. In the Top or Left viewport, move the omni light upward so it is approximately level with the spot light.

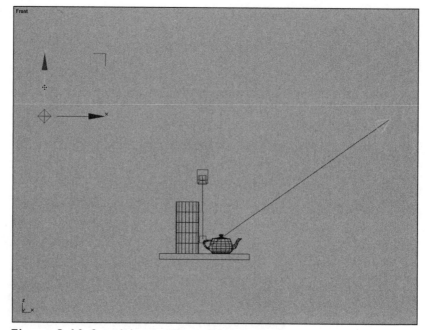

Figure 3.10 *Omni light moved up in Front viewport*

6. Render the Camera01 view.

The scene is now more evenly illuminated.

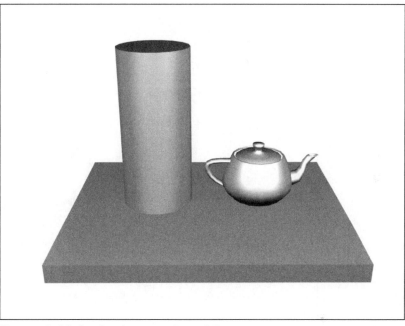

Figure 3.11 *Rendered scene with two lights*

Enabling Shadows

Next you'll make the spot light cast shadows.

1. Select the spot light.

2. Click **Modify** to access the Modify panel.

3. At the top of the General Parameters rollout, check the **Cast Shadows** check-box.

The viewport doesn't show the shadows. You'll need to render the scene to see the shadows.

4. Render the Camera01 view.

The scene now has shadows cast by the spot light.

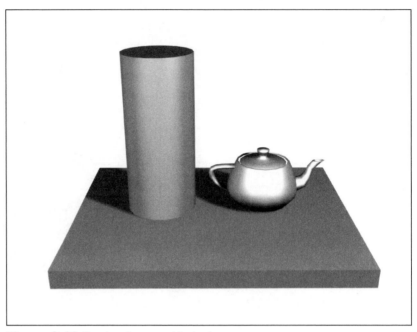

Figure 3.12 *Rendered scene with shadows*

Increasing Contrast

Next, you will reduce the intensity of the omni light to create more contrast in the scene. The light intensity is set with the light's V value on the command panel.

1. Select the omni light.

2. On the General Parameters rollout, locate the RGB and HSV settings. Change the **V** value to 100.

3. Render the Camera01 view.

There is now more contrast in the scene.

Figure 3.13 *Rendered scene with shadows and contrast*

Adjusting Hotspot and Falloff

Every direct and spot light has both a *hotspot* and *falloff* area.

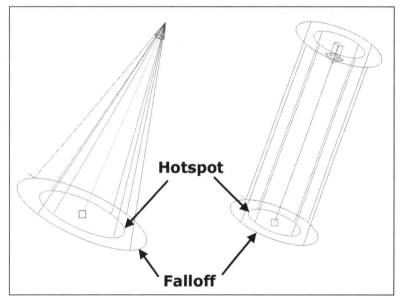

Figure 3.14 *Hotspot and falloff*

The hotspot determines the area where the light will shine at its full V value intensity. Between the hotspot and falloff, the light gradually "falls off" or decreases in intensity. The light has no effect outside the falloff area.

The hotspot and falloff are expressed as a number of degrees off the line from the light to its target. In practice, the hotspot and falloff are set visually rather than numerically.

1. Select the spot light.

2. On the Spotlight Parameters rollout on the command panel, locate the Hotspot and Falloff parameters.

3. Increase or decrease the **Hotspot** and **Falloff** values while watching the hotspot and falloff circles in the Camera01 view. Change the hotspot and falloff so they look similar to Figure 3.15.

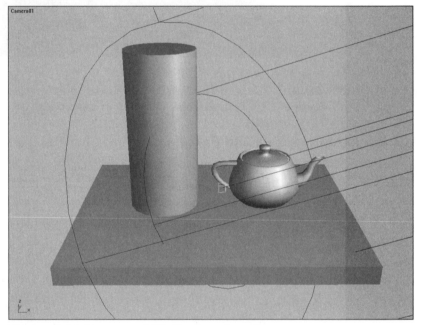

Figure 3.15 *Hotspot and falloff angles adjusted*

4. Render the Camera01 view.

Some portions of the table and objects are darker because of the new hotspot and falloff angles.

Figure 3.16 *Rendered image with adjusted hotspot and falloff angles*

Summary

In this Quick Start, you have learned that:

- A target light or camera is created by clicking and dragging from the light or camera source to the target.

- Lights and cameras are always placed on the construction plane, meaning they usually have to be moved in a different viewport in order to point in the appropriate direction.

- More than one light is usually necessary in a scene for realistic illumination and contrast.

- Light intensity is set with the V value on the command panel.

- Spot and direct light have a hotspot and falloff angle for determining the area where the light will fall.

- Although you can get an idea of how lights affect a scene by looking in a shaded viewport, the full effects of lights and shadows appear only in renderings. (An exception to this rule is the ActiveShade viewport. See *New Features in 3ds max 4* on page 57.)

- Shadows appear in a rendering only when the Cast Shadows checkbox has been checked for at least one light.

WORKFLOW

PLANNING

Before placing lights in a scene, consider the light sources you wish to represent. Where is the light coming from? Where should the shadows fall?

Draw a pencil sketch on paper of your scene from the Top viewport. Draw the lights and camera(s) in the sketch at their approximate locations, and set an intensity for each light corresponding to the V value on the command panel. As a rule of thumb, the total of all V values for each light should fall between 300 and 600. You will often change these intensities in the scene once you see the rendered image, but this guideline will give you a starting point.

You should also set the type of light to be used for each light source. The three types of lights in 3ds max 4 have different properties:

- Direct lights cast parallel shadows, making them the best choice for sunlight.

- Spot lights cast non-parallel shadows, making them suitable for indoor shadow-casting lights.

- Omni lights are best for overall illumination and for filling in dark spots. Avoid using omni lights to cast shadows.

For direct and spot lights, determine whether the light should have a wide or narrow hotspot and falloff. You can also choose between a rectangular or circular cone.

PLACING THE CAMERA

In the 3D scene, create the camera in the Top viewport, and move it into place in the Left and/or Front viewports. Change the Perspective view to the Camera view and adjust the camera as needed.

PLACING LIGHTS

Create lights in the Top viewport and move them into place in other viewports. When you place the first light, the scene will go dark in any shaded views such as the Perspective view. Disregard this problem and place the remaining lights. After moving the lights into place, render the camera view to see the effect of the the initial light placement, and adjust as needed.

RENDERING TESTS

It is not unusual to change the lighting in a scene a dozen or more times before it looks right. Most scenes require no more than three or four lights for basic illumination. Additional lights can be added for special effects.

If the basic illumination is not working out, work with the lights you have, or perhaps add one or two more. Don't keep adding lights when you can't even get the basic lights working. Extra lights increase rendering time. If a light seems to be having no effect, delete it or turn it off by unchecking the On checkbox on the General Parameters roll-out.

If your scene has numerous complex objects and textures, you can save a lot of time by creating a stripped-down version of your scene for your first lighting tests. Save your scene in its current state, then replace complex objects and textures with simple ones. When the lighting looks good, save the simpler scene, load the complex one and import the lights.

NEW FEATURES IN 3DS MAX 4

ACTIVESHADE VIEW

The ActiveShade view is a new type of viewport in **3ds max 4**. An ActiveShade viewport contains a preview rendering showing lights, shadows and materials. When you change the lighting or materials, the ActiveShade viewport updates instantly and automatically.

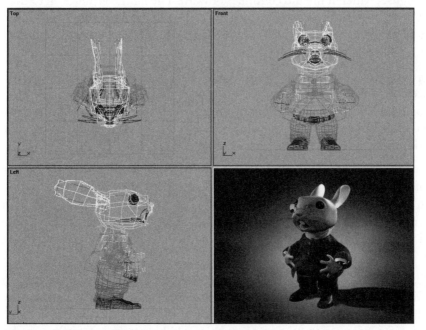

Figure 3.17 *ActiveShade view in lower right viewport*

An ActiveShade viewport can take the place of a standard viewport, or it can float free on the screen.

To change an existing viewport to an ActiveShade viewport, activate the viewport and choose *Rendering/ActiveShade Viewport* from the menu.

To create a new floating ActiveShade viewport, click the **ActiveShade Floater** button at the rightmost end of the Main Toolbar. You can also choose *Rendering/ActiveShade Floater* from the menu.

TIPS AND TECHNIQUES

Lighting a scene correctly requires some planning and ingenuity. Lights in **3ds max 4** don't simulate real life lights exactly, so you'll have to get creative to make your lighting look realistic.

LIGHTS IN LIFE

In life, you are accustomed to the general properties of light. You turn on a lamp, and the area near the lamp is strongly illuminated. Areas across the room aren't so well lit. Let's take a look at what's really going on here.

A light ray leaves a light source and continues in a straight line until it reaches an obstacle it cannot pass through, such as a solid object. It then bounces off the object at the same angle at which it hit and continues onward. The light continues on to the next solid object, losing a bit of energy with each bounce until it finally wears out. This bouncing effect allows light to go around corners, creep through cracks, and illuminate a floor under a table that isn't receiving direct rays of light.

Light intensity decreases as rays of light travel farther from the source. Light also bounces off airborne particles and even the air itself, diffusing the light rays and preventing them from bouncing infinitely in any one direction. These factors explain why a lamp illuminates the area near the light source more than it does the far end of the room. This diminishing effect is called *attenuation*.

When light hits an object, the object absorbs some of the colors and reflects others. When light reflects (bounces) off an object, it carries some of the object's color with it which in turn colors the next object it strikes. For example, an object placed on a smooth, shiny red table will be visibly tinted red in the areas that face the tabletop. This effect is called *radiosity*. With objects that aren't so shiny or brightly colored, radiosity still takes place, but the effect is much more subtle.

In life, light can also create reflection (light bouncing off a smooth surface) and refraction (light bending as it passes through transparent media such as water or glass). In **3ds max 4**, these properties are represented by materials applied to an object. See *Chapter 4: Materials* for information on how to make a 3D object appear to reflect or refract.

LIGHTS IN 3DS MAX 4

Light sources in **3ds max 4** don't behave exactly like lights in real life. With 3D lights, light rays continue infinitely in one direction regardless of obstacles encountered along the way. If shadows are enabled, the light rays stop when they encounter a 3D object but they never bounce. Radiosity isn't taken into account at all, but attenuation can be simulated with the settings on the light's Attenuation Parameters rollout.

The good news is that extra light sources can be placed in a 3D scene to simulate bounced light. In order to do this effectively, you have to be able to visualize how the lights would affect the scene in real life. This is where your sense of observation comes in. Noting various light setups in life and how they affect the objects around them is vital to becoming an expert at 3D lighting, and you can do it anytime, anywhere. Once you become a student of lighting in life, you'll find you can light your 3D scenes much more easily.

As an example, take a look at the following scene showing a cave, water, a bridge and a torch.

Figure 3.18 *Cave scene*

In the scene there is only one visible light source, the torch. If just one omni light is placed in the scene at the torch's location, the sides of the bridge posts facing the viewer are not illuminated.

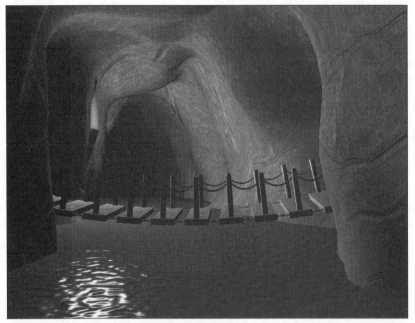

Figure 3.19 *Cave with one light*

In reality, the light from the torch would bounce off the walls to some degree and illuminate the sides of the bridge posts facing the viewer.

Figure 3.20 *Cave with additional light to simulate bounced light*

In Figure 3.20, a dim omni light has been placed on the side of the bridge closest to the viewer to simulate bounced light.

Some experimentation is always required to determine the exact lighting setup to do the job. Light types and intensities, excluded objects, and attenuation can all play a part in the simulation of bounced light.

RENDERING ASPECT RATIO

The *aspect ratio* is the relationship between an image or viewport's height and width. In order to see exactly what will appear in the final rendering, you'll need to set your Camera viewport to display the final rendering's aspect ratio.

Choose *Rendering/Render* from the menu, and on the Render Scene dialog, set the **Width** and **Height** to the final resolution you will be rendering. Close the dialog, then right-click the Camera viewport label and choose *Show Safe Frame* from the pop-up menu. This will crop parts of the viewport to display the output ratio, enabling you to see exactly what the rendered image will show when the Camera viewport is rendered.

For more information, see the *Rendering Scenes* topic on the ***3ds max 4 Interactive Quick Reference CD***.

SCENE COMPOSITION

In life, photographers and cinematographers use a variety of tricks to stage a scene. An extra light is placed to make a character or object the focus of a scene, while other areas are left dark. Cameras can be angled for more interesting and compelling composition. You can use these same techniques in your 3D work to bring your scenes to life.

The best way to learn these tricks is to study professional photographic and cinematic lighting and camera techniques. Books and magazines on these subjects can provide a wealth of knowledge that you can directly apply to your 3D scenes. Magazines and artwork showing the results of staged lighting and cameras can also give you a few ideas.

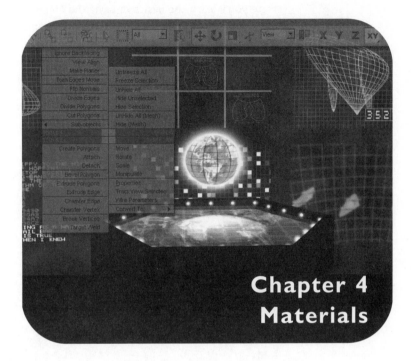

**Chapter 4
Materials**

The Material Editor is used to create materials and assign them to objects. Materials can be likened to paint or wallpaper placed on objects, setting the color or pattern to appear on an object. In addition, materials set the bumpiness, shininess, transparency and other aspects of the object to which they are applied.

The Material Editor is also used to set up maps for backgrounds and atmospheric effects.

Here you'll learn the basics of using the Material Editor. We'll start with the simple task of choosing and assigning a premade material and move on to custom material creation.

Scene Setup

1. Reset **3ds max 4**.

2. Create a sphere, a box and a teapot of similar size. Place them side by side as viewed in the Front viewport.

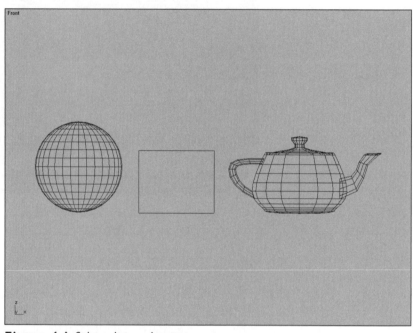

Figure 4.1 *Sphere, box and teapot*

These three objects were chosen for their varying shapes. The appearance of a material on an object depends largely on whether it's curved or flat.

Loading a Material

 I. Click the **Material Editor** button on the Main Toolbar to access the Material Editor.

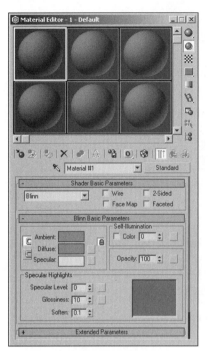

Figure 4.2 *Material Editor*

The Material Editor can be accessed in any of three ways:

- By clicking the **Material Editor** button on the Main Toolbar
- By choosing *Rendering/Material Editor* from the menu
- By pressing the **<M>** key on the keyboard

The six slots at the top of the Material Editor are called *sample slots*. The first sample slot is selected by default.

2. On the Material Editor window, locate the Get Material button.

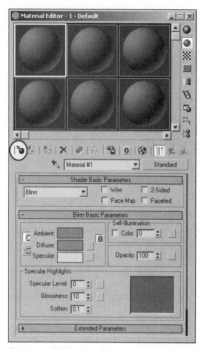

Figure 4.3 *Get Material button*

 3. Click **Get Material.**

The Material/Map Browser appears.

Figure 4.4 *Material/Map Browser*

4. On the Material/Map Browser, under the Browse From section at the left of the window, choose the **Mtl Library** option.

A list of premade maps and materials appears in the Material/Map Browser.

Figure 4.5 *Premade maps and materials*

 5. At the top of the Material/Map Browser window, click **View Large Icons**.

Wait a few moments as maps and materials are displayed as large thumbnails.

Figure 4.6 *Materials displayed as large icons*

6. Scroll through the maps and materials.

Note that the first ten thumbnails show flat pictures while later thumbnails display a sphere. The first ten thumbnail names are preceded by a green diamond to indicate that they are *maps*, while the sphere thumbnail names are preceded by a blue sphere to designate *materials*.

Maps and materials are not the same thing. A map is a 2D picture or pattern with no shininess or bumpiness. A material, on the other hand, can utilize several maps to define its color, shininess and bumpiness. Maps are 2D components while materials are 3D.

A map must be made part of a material definition before it can be applied to an object. Otherwise, **3ds max 4** wouldn't know how to use the map with the object.

7. Scroll through the Material/Map Browser and locate the material Bricks_Yellow. Double click the material image in the Material/Map Browser.

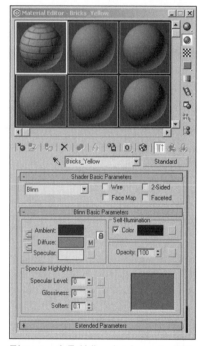

Figure 4.7 *Yellow bricks material in sample slot*

The material you just chose appears in the Material Editor in the first slot at the upper left.

8. Close the Material/Map Browser by clicking the **X** at its upper right corner.

Assigning the Material to an Object

1. Select the sphere object in your scene.

2. Click and drag the material from the Material Editor to the sphere.

   The material appears on the sphere in the Perspective view.

Figure 4.8 *Yellow brick material on sphere*

On the Material Editor, small triangles appear at the corners of the sample slot in to indicate that the material has been assigned to an object in the scene.

3. Activate the **Perspective** view.

4. Click **Quick Render (Production)** on the Main Toolbar.

   The material appears on the object in the rendering.

5. Click on the second sample slot on the Material Editor.

6. Get other materials from the Material/Map Browser and assign them to the box and teapot. Render the scene frequently to see how the different materials look. Use as many sample slots as you like to choose a variety of materials.

Creating Custom Materials

One of the great pleasures of **3ds max 4** is constructing your own custom materials. It's easy to spend many creative hours mixing and matching maps and colors to get that perfect look.

1. Select an unused sample slot. An unused sample slot has no triangles at the corners of its border, and displays a gray sphere.

2. Locate the Shader Basic Parameters rollout on the Material Editor window.

> ☆ **TIP** ☆
>
> *Most shaders, like Blinn, are named after the person(s) who developed them.*

By default, the shader is set to Blinn. A shader is simply a computation used to make a computer-generated object look like a real shaded (colored) object. The Blinn shader is used as the default because it renders quickly and is sufficient for most materials.

3. Locate the Blinn Basic Parameters rollout on the Material Editor.

This rollout contains three large color swatches labeled Ambient, Diffuse and Specular.

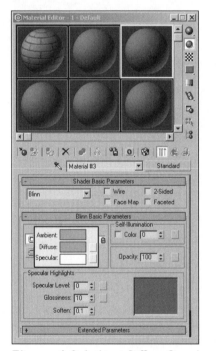

Figure 4.9 *Ambient, Diffuse, Specular color swatches*

- Ambient sets the color of the object for areas that are not well illuminated.

- Diffuse sets the main color of the object. This is the setting you will use most often.

- Specular sets the color of any highlights on a shiny object.

The Diffuse setting is the most important. This is where you determine the overall color of the object.

4. Click the Diffuse color swatch.

   A Color Selector appears.

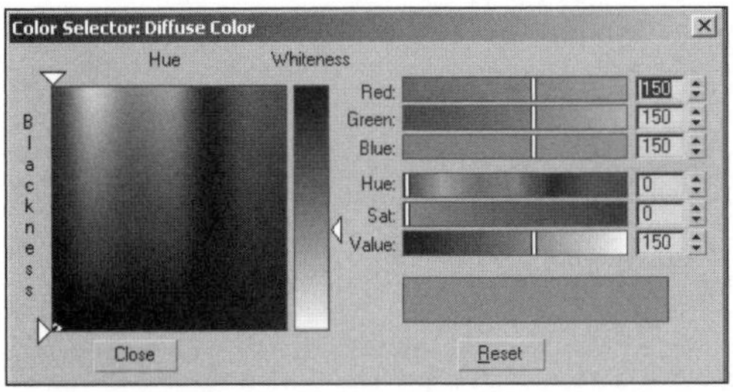

Figure 4.10 *Color Selector*

Note: If the Material/Map Browser appears, you have clicked in the wrong place. Close the Material/Map Browser and try again.

5. Use the sliders and/or numbers to change the color. As you change the color, the color of the sphere in the sample slot also changes.

6. When you have finished changing the color, close the Color Selector.

7. Assign the material to the sphere in the scene by dragging it from the sample slot to the sphere in the scene.

   The sphere in the scene changes color accordingly. A straight Diffuse color change to a material will always show in shaded viewports, while maps may or may not appear in viewports.

   Note that the sample slot in the Material Editor has small triangles at its corners to indicate the material has been assigned to an object in the scene.

Creating a Custom Material with a Map

So far you have only changed the Diffuse color. You can also assign a pattern or picture to the Diffuse attribute to change its overall colors. This is accomplished by assigning a map to the Diffuse attribute.

1. Locate the small empty box just to the right of the Diffuse color swatch. Click this box.

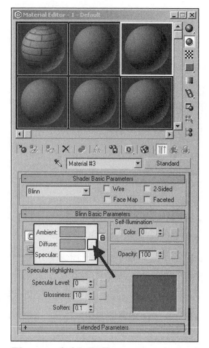

Figure 4.11 *Diffuse map selector box*

The Material/Map Browser appears.

2. Make sure the **New** option is selected under Browse From before continuing.

The Material/Map Browser displays different kinds of maps you can use to define the Diffuse attribute of the material. Here we'll use a bitmap, which is a 2D image such as a scan of a photograph.

3. Double-click on **Bitmap**.

A dialog appears where you can select a bitmap file.

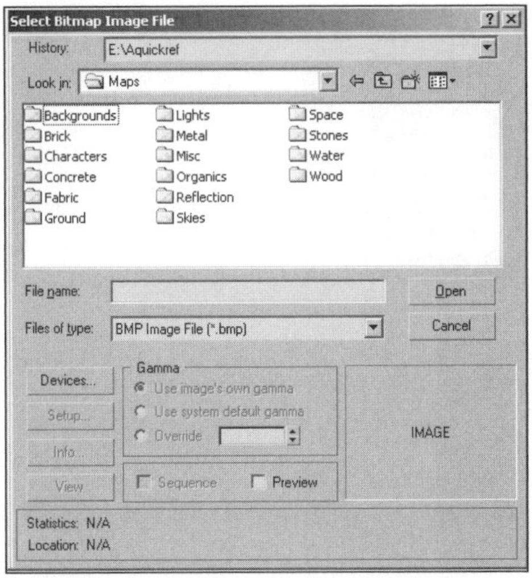

Figure 4.12 *File selector*

4. Check the **Preview** checkbox at the bottom center of the dialog.

5. Navigate to the *3dsmax4/Maps/Stones* folder, and click once on **lstone2.jpg**.

   A preview of this bitmap appears at the lower right of the dialog.

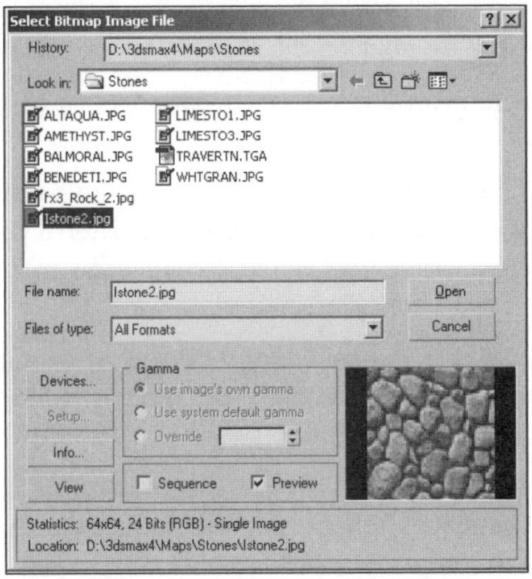

Figure 4.13 *Selected bitmap displayed on file selector*

6. Double-click **lstone2.jpg**.

   The sphere in the sample slot is now covered with the pattern shown in this bitmap.

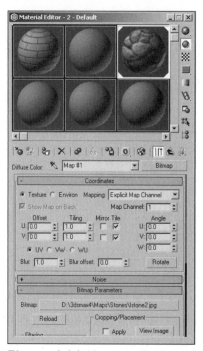

Figure 4.14 *Material with Istone2.jpg map in sample slot*

7. Click **Show Map in Viewport** on the Material Editor toolbar.

This will cause the stone map to appear in viewports on objects the material is assigned to.

The material is automatically updated on any object it was applied to previously. In addition, something else has also happened -- The Material Editor has changed! The Blinn Basic Parameters dialog has been replaced by new rollouts such as Coordinates, Noise and Bitmap Parameters.

This happens because the Material Editor works with a parent/child structure. The Diffuse map Istone2.jpg is a *child* of the entire material, while the material itself is the *parent*.

Changing any of the settings displayed will change the map (child). However, you will often want to get back to the parent level to make further changes to the material.

Working with Parent and Child Levels

1. Locate the **Go to Parent** button on the Material Editor toolbar. Click this button.

The Material Editor now displays the Blinn Basic Parameters rollout again.

2. Expand the Maps rollout on the Material Editor. You might have to scroll down the window to find this rollout.

TIP

You can assign a map to any attribute by clicking the None button in the Maps column. In the case of the Diffuse color and other attributes that appear on the Blinn Basic Parameters rollout, you can also click the small box next to the attribute name.

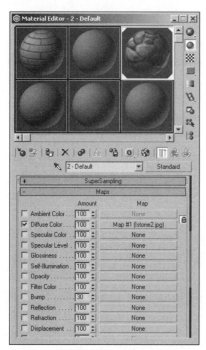

Figure 4.15 *Maps rollout*

The map lstone2.jpg is listed under the Map column next to Diffuse Color.

3. Click the button across from Diffuse labeled **lstone2.jpg**.

   The child level of the material is displayed again.

   You can also use the Material/Map Navigator to quickly move between child and parent levels,

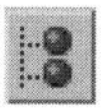

4. On the Material Editor toolbar, click **Material/Map Navigator**.

   The Material/Map Navigator appears.

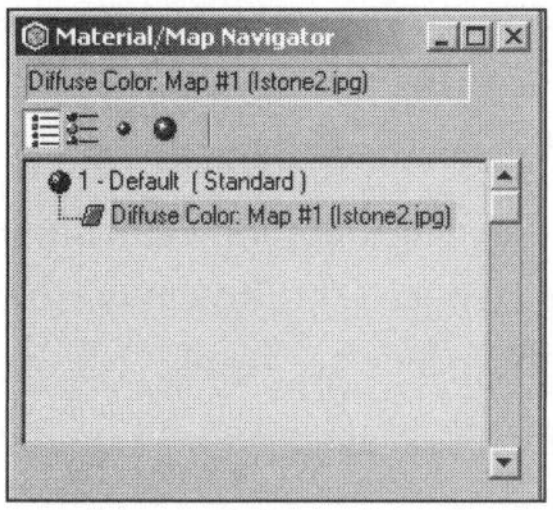

Figure 4.16 *Material/Map Navigator*

5. Click either of the listings on the Material/Map Navigator to move from one level to another.

Mapping Coordinates

When you assign a material containing maps to an object, **3ds max 4** needs to know how to lay the maps on the object. This is determined by the object's *mapping coordinates*.

Upon creation, most objects receive default mapping coordinates appropriate to their overall shapes. You can also assign new mapping coordinates to an object with the UVW Map modifier.

Here you'll assign new mapping coordinates to the objects in your scene and see the effects they create.

1. Select the stone material in the Material Editor.

2. Assign the stone material to all three objects in the scene.

Figure 4.17 *Stone material assigned to objects*

By default, the sphere is assigned Spherical mapping coordinates, meaning the stone image is stretched around the sphere. The box has Box mapping coordinates which places the image on each of the box's six faces. The teapot has separate mapping coordinates for the body of the teapot and its lid, handle and spout.

3. Select the sphere.

4. Go to the **Modify** panel.

5. From the **Modifier List** pull-down list, choose the **UVW Map** modifier.

6. On the command panel, select the **Box** option.

On the sphere, you can see the seam where the faces of the box map meet.

7. Select the **Spherical** option.

The sphere's mapping coordinates return to their original state.

Tiling

Currently the stone image is stretched across the sphere. To prevent stretching, the image can be tiled across the object. Tiling places more copies of the image on the object.

1. On the command panel, change **U Tile** and **V Tile** to 4.

The stone image repeats around the sphere.

Figure 4.18 *Tiled stone image on sphere*

2. Continue to experiment with maps, mapping coordinates and materials with the three objects in this scene.

Summary

We have just scratched the surface of the Material Editor here. In this Quick Start, you have learned that:

- Materials are created, edited and assigned on the Material Editor.
- Materials can be made up of maps.
- Maps cannot be assigned directly to objects, but materials made up of maps can.
- The Material Editor works on a parent/child system, where the material is the parent and all maps are children.
- You can navigate between parent and child levels with the Material/Map Navigator.
- Maps appear on objects in viewports only when **Show Map in Viewport** is checked on the Material Editor toolbar.

WORKFLOW

WHICH MAPS?

The first step in working with materials is to determine which maps you will use to define materials. Some of the maps will need to be scanned from photographs or drawn from scratch in a paint program such as Photoshop. You may even need to go out and take some new photographs.

DETERMINE MAPPING COORDINATES

Give some thought to the mapping coordinates you will use on each object. Sometimes the modeling approach is determined partially by the mapping coordinates you want the object to have in the end. If you plan to bend or deform an object, for example, you will probably want to assign mapping coordinates first.

MATERIAL CREATION

Create the materials in the Material Editor, and look at the sample slot to see how the material is shaping up.

The **Show End Result** toggle is very useful for alternately viewing individual maps and the entire material. Use it to check if individual maps are being utilized correctly.

The sample slot only gives an approximation of the material -- the real test is how the material looks in renderings. For this reason, you shouldn't spend a great deal of time fine-tuning the material at this stage.

APPLY MATERIALS

Apply the materials to the appropriate objects. You can do this by clicking and dragging from the sample slot to the object, or by selecting the object and clicking **Assign Material to Selection**.

If none of the material maps show up on the object in viewports, click **Show Map in Viewport** on the Material Editor toolbar.

The sample slots and viewports only gives an approximation of the material -- the real test is how the material looks in renderings. For this reason, you shouldn't spend a great deal of time fine-tuning the material at this stage.

RENDERING TESTS

Render the scene and adjust materials according to the rendering, not viewports or sample slots.

NEW FEATURES IN 3DS MAX 4

A Bricks map has also been added to **3ds max 4**. This map can be used to make custom bricks very quickly in any size or color.

The Combustion map replaces the Paint map to allow integration with Discreet's Combustion software. Combustion performs compositing and editing on video sequences, and is designed to work seamlessly with **3ds max 4**. The Combustion map in **3ds max 4** works only if you have installed the Combustion software on your computer.

TIPS AND TECHNIQUES

NAMING MATERIALS

Material names must be unique within the scene. For example, there can be only one instance of a material named Glass assigned to objects in the scene. However, you can have several materials with the same name in the Material Editor.

By default, each material is assigned a name such as **1 - Default**. You can change the name by typing in a new name on the Material Editor toolbar.

Keeping materials named intelligently is an important part of maintaining your scene, especially when the scene contains many objects.

REFLECTION

When you're standing in an illuminated room, some of the light bounces off objects and goes right toward you, enabling you to see the object. This same principle explains how mirrors and reflective surfaces work. Light bounces off an object, then off the mirror and into your eyes, enabling you to see the object in the mirrored surface.

The light bounces off the mirror at precisely the same angle at which it hits. You (or the camera) can only see the reflected object if positioned at the right spot.

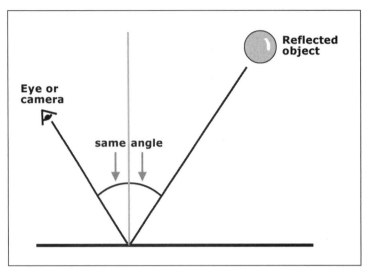

Figure 4.19 *Light bounces off mirror at the same angle it hits*

This is also why rough surfaces such as stucco or brick don't make good mirrors. Because of the roughness of the surface, light bounces off these surfaces in a variety of directions.

In **3ds max 4**, reflections are simulated with materials. Applying maps to the Reflection attribute on the Maps rollout will simulate a reflection.

- Using a Bitmap causes the object to reflect the bitmap in its surface.

- Using a Reflect/Refract or Raytrace map causes the object to reflect other objects in the scene. Raytrace is somewhat more accurate than Reflect/Refract.

Adding reflection to a scene increases its rendering time.

REFRACTION

When light hits certain materials such as glass or water it continues to move, but at a much slower rate. The slowdown causes the light to bend until it reaches the end of the obstacle, then it continues on in a straight line. This phenomenon, called *refraction*, causes an object behind the glass or water to appear bent, crooked or skewed.

Figure 4.20 *Refracting glass*

Refraction is simulated by assigning a map to the material's Refraction attribute on the Maps rollout.

- Using a Reflect/Refract map creates decent refraction under specific circumstances, and renders quite quickly.

- Using a Raytrace map creates very accurate refraction but takes a significantly longer time to render.

Adding refraction to a scene increases its rendering time even more than reflection does. Use this feature only when necessary.

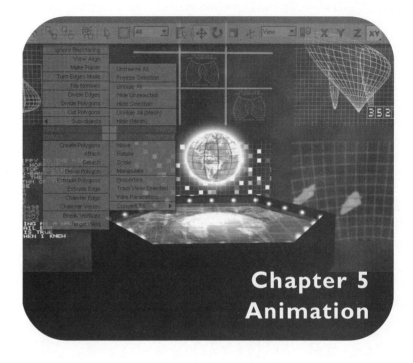

Chapter 5
Animation

One of the most significant advancements in the field of animation was the development of computer animation tools such as **3ds max 4**. Formerly, each frame of animation had to be drawn by hand by artists who trained for years at their craft. With animation software, you can learn to animate much more quickly as the results of your efforts can be viewed immediately on the screen.

In **3ds max 4**, animation is accomplished with the use of *keyframes*. You set object position, rotation and, and it figures out the in-b/or scale at specific frames, and **3ds max 4** figures out what the objects will do in between. This approach parallels traditional animation, where senior animators draw the key (important) frames, and junior animators draw the in-betweens. Here, you're the senior animator and **3ds max 4** is the junior.

QUICK START

This tutorial will get you started with animation in 3ds max 4.

In this exercise, you'll make a ball bounce repeatedly with looping, using Track View to adjust the animation. First you'll make the ball bounce once, then you'll use looping to make it bounce repeatedly.

Setting up the Scene

1. Reset **3ds max 4**.

2. In the Top viewport, create a sphere of any size.

3. In the Top viewport, create a flat box as a floor for the ball to bounce on.

4. In the Front viewport, move the ball so it sits a short distance above the box.

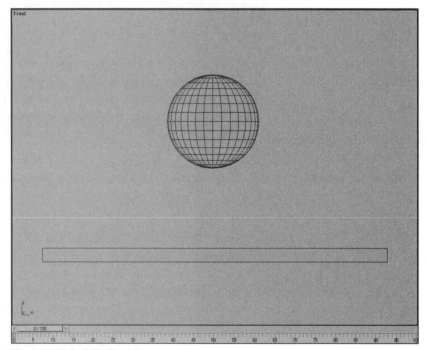

Figure 5.1 *Ball above box*

Animating the Ball

Animation keyframes are set with the help of the Animate button at the lower right of the screen. When clicked, this button turns red to show that it is in animation mode, and keyframes are set as you make changes to objects.

To move from one frame to another, you can click and drag on the time slider at the bottom of the viewport area until the desired frame is reached. You can also type in the frame number directly in the Animation Controls area at the bottom of the screen.

1. Go to frame 12.

2. Click the **Animate** button to turn it on.

3. In the Front viewport, move the ball down so it just touches the box.

You have just created a keyframe for the ball.

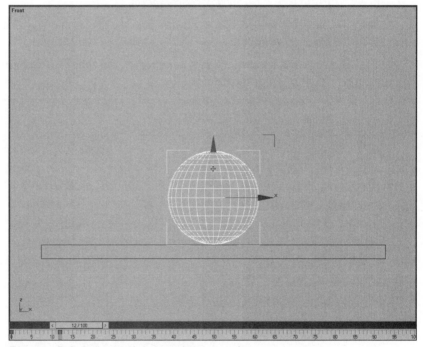

Figure 5.2 *Ball moved down to box*

Squashing the Ball

To make a bouncing ball look realistic, it needs to squash a little when it lands.

1. Go to frame 18.

2. On the Main Toolbar, choose **Select and Squash** from the **Select and Uniform Scale** flyout.
   You might see the following warning message:

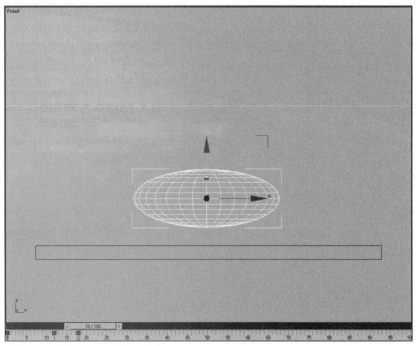

Figure 5.3 *Non-uniform scale warning*

Click **Yes** to continue. If you see this message in the future, simply click **Yes** to continue each time.

3. In the Front viewport, click and drag on the sphere's Y axis to squash it to about 2/3 of its original height.

Figure 5.4 *Squashed sphere*

The sphere squashes from the center. We would prefer that the sphere squash from the bottom.

Adjusting the Squash

1. Click **Undo** to undo the squash.

2. Turn off the **Animate** button.

3. Go to the **Modify** panel. The parameters for the sphere appear. Check the **Base to Pivot** checkbox.

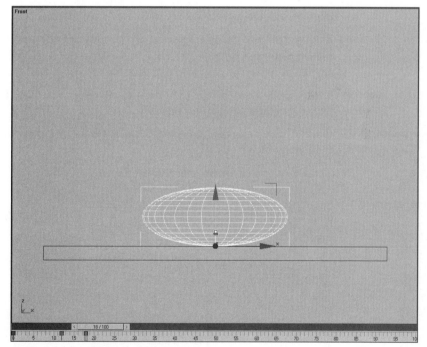

Checking **Base to Pivot** moves the sphere's pivot point to the bottom of the sphere. The pivot point is the point around which all transforms take place. Changing the pivot point will also affect the position of the sphere on frame 12.

4. Go to frame 12.

5. Turn on the **Animate** button.

6. Move the sphere again so it just touches the box.

7. Go back to frame 18. Use **Select and Squash** again to squash the sphere to about 2/3 of its original height.

Figure 5.5 *Ball squashed from bottom*

8. Turn off the **Animate** button.

Preparing Track View

You now have all the keys you need for the bouncing animation. Making the ball return to its original scale and position is simply a matter of copying keys in Track View.

1. On the Main Toobar, click **Track View** to open a new Track View window.

Figure 5.6 *Track View window*

The Track View window consists of toolbar buttons and a hierarchy listing down the left side. Each item in the hierarchy has a track on the right. When a track has animation set up for it, key dots appear for each keyframe in the track area.

2. On the Track View toolbar, right-click the **Filters** button and select *Animated Tracks Only* from the pop-up menu.

3. On the Track View hierarchy, right-click on **Objects** and choose *Expand All* from the pop-up menu.

These two actions cause only animated tracks to appear in the hierarchy. The key dots for each animated track are displayed.

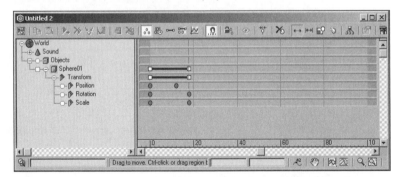

Figure 5.7 *Animated tracks only*

TIP: The Select and Squash function creates Scale keys in Track View. Rotation keys are sometimes created even though no rotation has taken place. We will ignore the rotation keys and the animation will still work out fine.

Copying Keys

Getting the ball to bounce up and down is simply a matter of copying the appropriate keys in Track View. To copy a key, make sure the **Move Keys** option on the Track View toolbar is selected. Hold down the **<Shift>** key and click and drag the key dot in the Track View track display. You can see which frame you're copying to by watching the frame number display at the bottom of the Track View window.

1. Copy the key dot on the Position track at frame 12 to frame 24.

   This will make the ball stay on the box until it's ready to come back up.

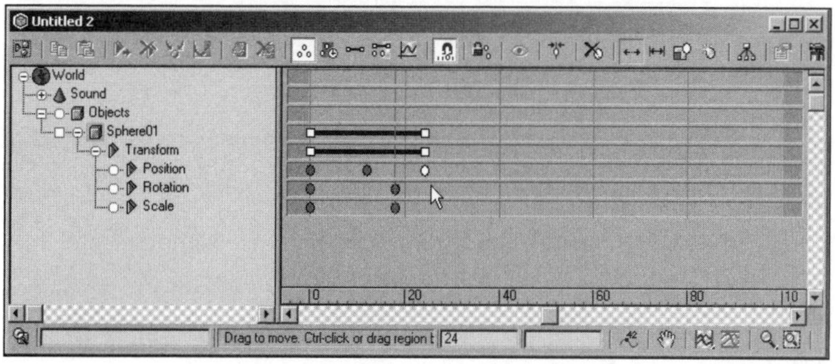

Figure 5.8 *Position key copied from frame 12 to 24*

2. Copy the key dot on the Position track at frame 0 to frame 36.

   This will make the ball move back up to its original position after the bounce.

3. Copy the key dot on the Scale track at frame 0 to frame 12.

   This will keep the ball at its original scale until it's ready to squash.

4. Copy the key dot on the Scale track at frame 0 to frame 24.

   This will return the ball to its original scale after the squash.

5. Copy the key dot on the Scale track at frame 0 to frame 36.

   This puts the last Scale key at the same frame as the last Position key, so looping will work later on.

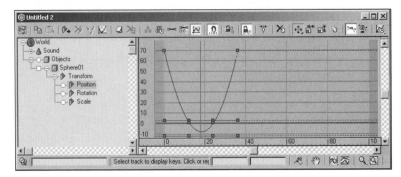

Figure 5.9 *Track View after keys have been copied*

6. Click **Play Animation** to view the single bounce.

The ball moves down, squashes, then returns to its original position.

The animation has a few problems:

The ball passes through the box between frames 12 and 24.

These problems can all be fixed in Track View by using *function curves*.

Working with Function Curves

Function curves show your animation as line graphs. This makes it easy to see where a track dips below a certain value. Function curves can also be used to adjust the animation.

1. Highlight the **Position** track in the hierarchy listing.

2. On the Track View toolbar, click **Function Curves**.

The function curves for the Position track appear.

<table>
<tr><td>

☀ TIP ☀

*If you can't see the entire graph, click **Zoom Value Extents** at the bottom right of the Track View window.*

</td></tr>
</table>

Figure 5.10 *Function curves for Position track*

Each keyframe is displayed on the graph as a dot. Here you can see that the graph dips below the 0 mark between frames 12 and 24. This is because by default, **3ds max 4** always attempts to make a smooth transition from one keyframe to another. We want the line to be straight between frames 12 and 24 so the ball will stay on the box and not pass through it while it's squashing and unsquashing before moving back up from the bounce.

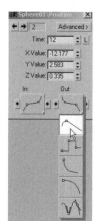

3. Right-click any key at frame 12 to display the **Sphere01\Position** pop-up window.

4. Click and hold the **Out** button. Choose the second button displayed, which shows a straight line coming out from a key dot.

5. On the **Sphere01\Position** window, click the right arrow to move to the third keyframe at frame 24.

6. Click and hold the **In** button. Choose the second button displayed.

7. Close the **Sphere01\Position** window.

   The function curve between frames 12 and 24 is now straight. This happened because we set both the outgoing curve on frame 12 and the incoming curve on frame 24 to straight lines.

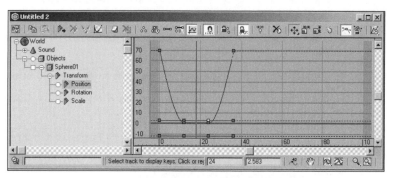

Figure 5.11 *Straight line between frames 12 and 24*

8. Play the animation again.

   The ball no longer moves through the table between frames 12 and 24. However, the ball appears to stretch as it moves toward the box, then stretches again as it moves back up. This is because of the function curves created when the Scale keys were set.

Adjusting the Scale Function Curves

1. Highlight the **Scale** track in the hierarchy listing.

   The Scale curves appear.

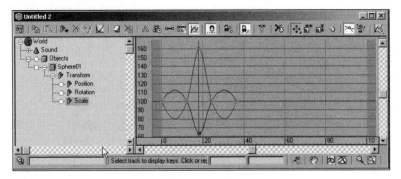

Figure 5.12 *Function curves for Scale track*

We'd like the scale to remain the same between 0 and 12 and between 24 and 36. From the graph you can see this is not so.

2. Right-click any key at frame 0 to display the **Sphere01\Scale** window. Click and hold the **Out** button and choose the second button listed.

3. Click the right arrow on the **Sphere01\Scale** window to move to key 2 on frame 12.

4. Click and hold the In button to change it to the second button.

5. Click the right arrow twice move to key 4 on frame 24, Change the **Out** button to the second choice.

6. Click the right arrow to move to key 5 on frame 36. Change the In button to the second choice.

7. Close the **Sphere01\Scale** window.

   The function curves between frames 0 and 12 and between frames 24 and 36 are now straight.

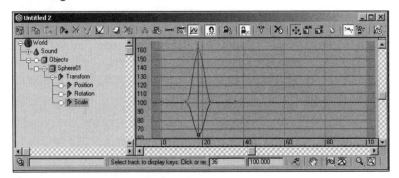

Figure 5.13 *Adjusted function curves for Scale track*

Play the animation to check your work. The ball should no longer stretch as it descends or ascends.

Looping the Animation

Now you're nearly ready to loop the animation. The first consideration is the length of the animation. The bounce occurs over 36 frames, so to make a smoothly looping animation, the total number of frames should be a multiple of 36. We'll set the frame count to 108, which is 36 times 3.

 1. Click the **Time Configuration** button in the Animation Controls area at the lower right of the screen.

2. On the Time Configuration dialog, set the **Frame Count** parameter to 108. Click **OK** to close the dialog.

Next, you'll make the Position and Scale tracks loop.

3. On the Track View hierarchy, highlight both the **Scale** and **Position** tracks by highlighting one, then holding down the **<Ctrl>** key and clicking the other.

4. On the Track View toolbar, click **Parameter Curve Out-of-Range Types**.

The Param Curves Out-of-Range Types dialog appears.

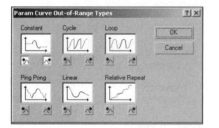

Figure 5.14 *Param Curves Out-of-Range dialog*

5. On the Param Curves Out-of-Range Types dialog, click the graphic picture under **Loop**. Click **OK** to close the dialog.

Play the animation. The ball bounces up and down repeatedly. However, it appears to hit something hard when it reaches the top of its travels. You'll use function curves again to fix this problem.

Adjusting the Looped Bounce

1. On the Track View hierarchy, highlight the Position track to display just the function curves for this track only.

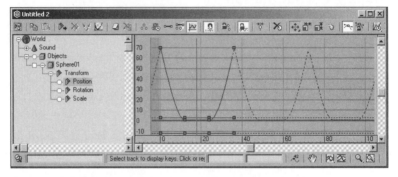

Figure 5.15 *Function curves for looped Position track*

Note that the curve arches down sharply as it leaves frame 0, then arches up again to frame 36. The function curves continue before and after frames 0 and 36 as a dotted line. You can see that there is an abrupt change in the curve at the beginning and end of the looped sequence. This is what causes the ball to make such an abrupt reverse in direction at that time.

To make the ball traverse smoothly, you'll change the function curves going out of frame 0 and coming into frame 36. This will change the curves throughout the entire loop.

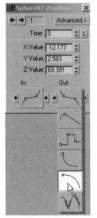

2. Right-click any key on frame 0 to display the **Sphere01/Position** dialog. Click and hold on the **Out** button, and choose the second-to-last selection on the flyout, which shows the curve making a bell shape as it comes out of the dot.

3. Click on any key at frame 36 to select key 4 at frame 36. Click and hold the In button, and choose the second-to-last selection from the flyout.

Your function curves should similar to the following figure.

Figure 5.16 *Function curve for smooth looping*

4. Close Track View.

5. Play the animation.

The ball bounces smoothly throughout the animation.

Linking

Now that you've learned to work with keyframes and function curves, we'll have a little fun with linked objects and the Flex modifier.

1. Turn off the **Animate** button.

2. Go to frame 0.

3. In the Top viewport, create a small cylinder. Move the cylinder so it sits slightly inside the top of the ball.

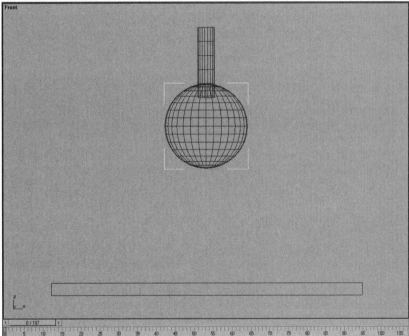

☼ TIP ☼

If the cylinder is the wrong size, don't scale it with transforms. Use the Modify panel to change its Height or Radius.

Figure 5.17 *Cylinder placed on sphere*

4. In the Front viewport, rotate the cylinder to approximately a 45 degree angle on the Z axis.

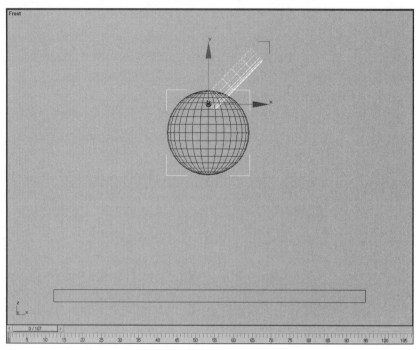

Figure 5.18 *Cylinder rotated*

5. Copy and rotate the cylinder to the other side of the ball by holding down the **<Shift>** while rotating on the Z axis. On the Clone Options dialog, choose the **Copy** option and click **OK.**

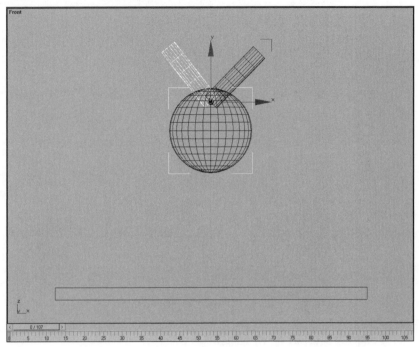

Figure 5.19 *Copied cylinder*

6. Select both cylinders.

7. Click **Select and Link** on the Main Toolbar.

8. Click and drag from the cylinders to the ball.

   The cylinders are now linked to the ball.

9. Play the animation.

The cylinders move along with the ball, and squash when the ball squashes.

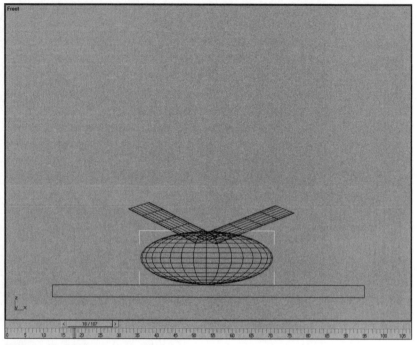

Figure 5.20 *Squashing cylinders*

Applying Flex to the Cylinders

The cylinders can be made to bend and flex with each bounce with the Flex modifier.

1. Select both cylinders.

 2. Go to the **Modify** panel.

3. Click the **Modifier List** pulldown arrow, and press the **<F>** key until the Flex modifier is selected. Click **Flex**.

The Flex modifier has been applied to the cylinders.

4. Play the animation.

The cylinders now flop around as the ball bounces.

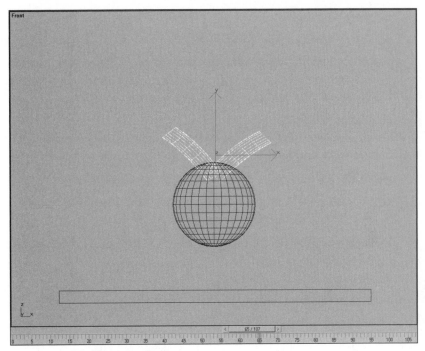

Figure 5.21 *Flopping cylinders*

Summary

In this Quick Start, you have learned that:

- Animation is set up with keyframes at key points in the animation.

- Keyframes are set by turning on the Animate button and transforming the object.

- Track View is used to copy keys, adjust the motion and loop the animation.

- Objects can be linked together to follow one another during the animation.

- The Flex modifier causes linked child objects to react to motion.

WORKFLOW

PLAN YOUR ANIMATION

Before creating objects, consider the structure of your animation. Figure out which objects will remain stationary and which will be animated. Animated objects can be linked together where appropriate to make your job easier.

Animated objects should never be grouped with the *Group/Group* menu option. Grouping and ungrouping animated objects can give unpredictable results. Instead, use linking with animated objects.

CREATE OBJECTS

Objects that will be squashed or bent must have a sufficient number of segments or faces to allow the deformation to take place.

When objects are linked together, it's important that the pivot points be placed in the right locations.

Avoid using the scale transforms on objects you plan to link to other objects and animate with rotation. Scaling with transforms can cause a linked object to skew wildly in the animation. The animation shown in the *Quick Start* used a scale transform only on the parent object and no rotation in the child objects, so this problem did not occur. In practice, you should use other methods such as the Xform modifier to scale objects during animation. See the *Linking* topic on the **3ds max 4 Interactive Quick Reference CD** for more information.

Instead of scaling linked objects, use parameters to change the size. For example, if you want to make a box smaller, change the box's Length, Width or Height parameters on the Modify panel.

LINK OBJECTS AND TEST LINKAGES

When objects are linked together, the object being linked is called the *child*, while the object it is linked to is called the *parent*. When the parent is transformed (moved, rotated, scaled) or animated, the child follows it. A parent can have more than one child, but a child can have only one parent. The main parent is called the *root*.

Be sure to link objects starting from the child and going to the parent. Test any links by moving the root object and making sure all the children go with it.

If you start animating and later want to add another link, be sure to go to a frame where the objects are in the right relationship to one another.

CREATE KEYFRAME ANIMATION

Turn on the **Animate** button and start moving objects around.

ADJUST KEYS IN TRACK VIEW

Any animation beyond the most basic requires some adjustment in Track View. Look at the function curves and adjust as necessary to make the animation work the way you want it to.

USE CONTROLLERS

Animation controllers are additional tools that go beyond keyframe animation in their capabilities. Controllers can be used to make an object follow a spline path or to force one object to point at another object. See the *Animation Controllers* topic on the *3ds max 4 Interactive Quick Reference CD* for more information.

NEW FEATURES IN 3DS MAX 4

FLEX MODIFIER

The Flex modifier has been enhanced in 3ds max 4 to allow deformation of objects with soft body dynamics. This feature detects collisions with other objects and responds with the appropriate squashing and stretching.

For information on how to use this new feature, see the *Flex* topic on the *3ds max 4 Interactive Quick Reference CD.*

SKIN MODIFIER

The Skin modifier deforms a mesh according to the animation of an underlying skeleton. Improvements to the Skin modifier in **3ds max 4** allows custom changes to specific areas of the skinned object based on the angle between two adjacent bones.

For example, when a character's elbow bends, the skin on the inside arm area might pinch or curve unnaturally. Rather than making endless changes to envelopes in an effort to solve the problem, you can simply specify that when two adjacent bones are at a certain angle to one another, the skin will react in a particular way. This feature can also be used to make muscle bulges when an elbow or knee is bent. The new tools are called *angle deformer*s.

For information on angle deformers, see the *Skin* topic on the *3ds max 4 Interactive Quick Reference CD.*

BONES

Enhancements to the Bones objects on the **Create / System** command panel make them easier to use than ever. Bones are much easier to move and copy, and can be changed in size and made renderable. New animation tools called *IK solvers* can be used with bones to quickly set up a working bone structure.

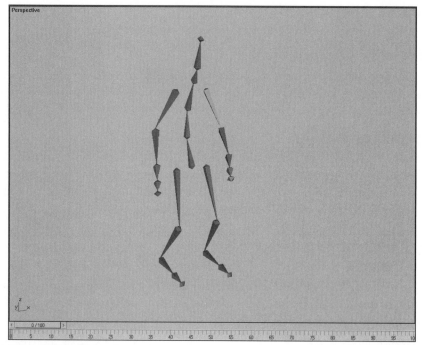

Figure 5.22 *New Bones*

For information on how to use Bones, see the *Bones* topic on the ***3ds max 4 Interactive Quick Reference CD***.

IK SOLVERS

In **3ds max 4**, you have two options for animating a linked chain: forward kinematics (FK) and inverse kinematics (IK). Forward kinematics is the default type of animation used in **3ds max 4**. If you move, rotate or scale a parent object, the child object(s) follow along with it.

Inverse kinematics works in a somewhat opposite manner. When a child object is moved, the parent objects all the way up the chain rotate to accommodate the new child position. The exception is the root (main parent) object, which stays still. This tool is very useful in character animation, where you can move a hand to make the entire arm move.

In **3ds max 4**, IK is enabled with the help of new tools called *IK solvers*. These tools set the start and end points where IK will be used in the chain. Each IK Solver creates a separate object called an *IK chain* that goes from the first to the last object designated to work with IK.

For information on how to set up and use IK solvers, see the *IK Solvers* topic on the ***3ds max 4 Interactive Quick Reference CD***.

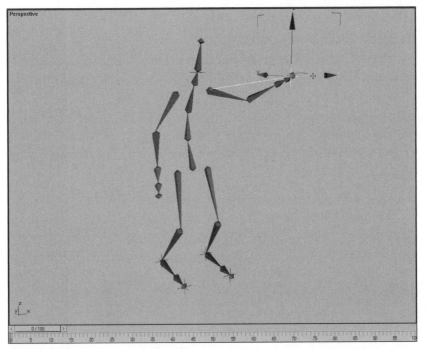

Figure 5.23 *Bones used with IK*

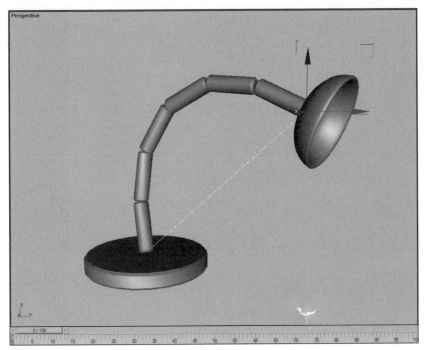

Figure 5.24 *IK solver used with lamp*

TIPS AND TECHNIQUES

WATCH THAT ANIMATE BUTTON!

It's easy to forget to turn the **Animate** button on or off. This problem plagues even long-time animators. There is no known cure for this deficiency, but good work habits can get around it.

Save your work frequently. That way, if you mess up your animation and the **Undo** button isn't enough, you can always return to a previously saved version. In fact, this is a good practice regardless of your personal relationship with the **Animate** button.

TRACK VIEW

If you want to create beautifully animated scenes, Track View is the most important tool to master. Moving objects around with keyframes isn't enough. You need to be able to make objects slow down, stop and start at the appropriate rate. This is possible only with Track View in general, and with the Function Curves feature in particular.

The *3ds max 4 Interactive Quick Reference* consists of a comprehensive help system covering every feature in **3ds max 4**. Explanations of every tool, option and parameter can be quickly accessed. Complete, clear descriptions of each parameter are listed with each option.

In addition, you can jump between related topics at the click of a button to enhance your understanding of any topic.

SYSTEM REQUIREMENTS

The following are required to install the *3ds max 4 Interactive Quick Reference CD*:

RECOMMENDED	MINIMUM
Intel Pentium processor II or compatible, 400MHz or faster	Intel Pentium processor or compatible, 120MHz or faster
64MB RAM	32MB RAM
Windows NT4, 2000, Me	Windows 95, 98, NT4, 2000, Me
60MB free disk space	50MB free disk space
3ds max 4 installed	**3ds max 4** installed
Internet Explorer 5 or higher	Internet Explorer 4

To download Internet Explorer, visit http://www.microsoft.com/ie.

INSTALLING THE CD

3ds max 4 must be installed on your computer before you can install the *3ds max 4 Interactive Quick Reference CD*.

To install the *3ds max 4 Interactive Quick Reference CD*:

1. Close any open programs.

2. Insert the **3ds max 4 Interactive Quick Reference CD** into your CD-ROM drive. After a few moments, the installation screen appears.

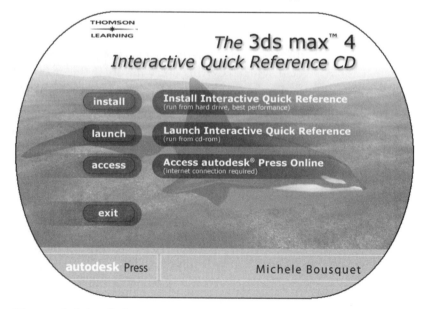

Figure A.1 *Installation screen*

If the installation screen doesn't appear, then you will need to run the installation routine manually:

- On the Windows desktop, double-click the **My Computer** icon.

- Double-click the icon for your CD-ROM drive in the My Computer window.

The installation screen will appear.

3. Choose one of the options on the installation screen:

- **Install** installs the *3ds max 4 Interactive Quick Reference* on your hard disk. The application will be available from the Windows Start menu, and from inside **3ds max 4**.

- **Launch** runs the *3ds max 4 Interactive Quick Reference* directly from the CD. You will have to have the CD in the CD-ROM drive at all times in order to access the application.

- **Access** goes directly to the Autodesk Press Online website where you can find updates, additional tutorials and other tools. You must be connected to the Internet to use this option.

ACCESSING THE 3DS MAX 4 INTERACTIVE QUICK REFERENCE

RUNNING THE INSTALLED APPLICATION

If you have chosen to install the application, you can run it from within **3ds max 4**. From the *Help* menu, choose *Additional Help*, then choose **3ds max 4 Quick Reference** from the Additional Help list.

You can also run the software as a stand-alone application. Click **Start**, then **Programs**, then **Autodesk Press**, then **3ds max 4 Quick Reference**.

LAUNCHING FROM THE CD

If you have chosen to launch the *3ds max 4 Interactive Quick Reference* from the CD, the user interface will appear immediately.

USING THE 3DS MAX 4 INTERACTIVE QUICK REFERENCE

The *3ds max 4 Interactive Quick Reference* is a Windows-style help system. The application launches in your default Internet browser.

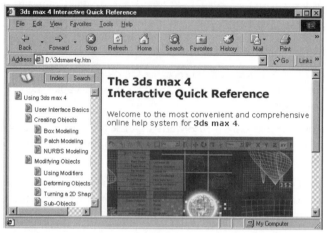

Figure A.2 *Interactive Quick Reference user interface*

The **Table of Contents** tab shows the list of topics in the *3ds max 4 Interactive Quick Reference*. Click on a topic to view it in the display window.

Use the **Index** tab to look for a particular option or parameter. Click on a topic to view it in the display window.

Use the **Search** tab to search all topics with keywords.

You can also use the buttons on your browser such as **Back** and **Forward** to move between topics.

TECHNICAL SUPPORT

If you encounter problems with installing or running the *3ds max 4 Interactive Quick Reference CD*, contact us by phone, fax or email.

Phone   (800) 477-3692 M-F 8:30am-5:30pm EST

Fax   (518) 464-7000

Email   help@delmar.com